# Soil Biology in Tropical Ecosystems

Tancredo Souza

# Soil Biology in Tropical Ecosystems

 Springer

Tancredo Souza
Centre for Functional Ecology
University of Coimbra
Coimbra, Portugal

ISBN 978-3-031-00966-2          ISBN 978-3-031-00949-5    (eBook)
https://doi.org/10.1007/978-3-031-00949-5

This Springer imprint is published by the registered company Springer Nature Switzerland AG
The registered company address is: Gewerbestrasse 11, 6330 Cham, Switzerland

*To my wife Fransuene*
*Without whom this book would have never*
*been completed. She helped me in more ways*
*than anyone else during dark days.*
*To my beloved mother, Rejane Souza*
*She truly was a force of nature. Her life was*
*a blessing, her memory a treasure.*
*Everything not saved will be lost (Nintendo*
*"Quit Screen" message).*

# Contents

# Preface

When I began studying soil ecology in the first year of my postdoc, I realized that it would be a tough task considering soil biology at the tropical ecosystem. There are plenty of references on the Internet that confuse the new student or new researcher in this area, and unfortunately, a lot of papers considering the tropical ecosystem are published in predatory journals. I only found a few books and some scientific papers about soil ecology that were written considering results obtained in the tropics, and that made me start writing this handbook.

Here, I made a compilation of several key points on soil ecology from the definition of soil to the influence of natural disasters on soil ecosystem. My focus is to provide you, a beginner, key information about soil ecology, how to study soil organisms, and how to assess the plant-soil interactions and their feedback. After reading this book, I hope you can start your field experiments with more knowledge and accuracy. Trust me, you will be able to do it.

In fact, you'll understand that studying soil organisms in the tropics is an adventure, because you will need to know many things about different scientific areas and make a network with them, such as (1) soil science; (2) soil biology; (3) soil ecology; (4) geography; (5) zoology; (6) bioassays and inoculation; and, finally, (7) statistical analysis. If you are thinking that it's a hard adventure, I tell you: It's true, but once you finish reading this book, you will be able to start your own experiments, everything will be all right. Don't forget that I went through the same; in the beginning it was hard, but now it's a pleasure to study soil biology. It's my work.

My focus is to provide you with key information, do you remember? This book is going to help you. So, it's a great pleasure for me to introduce you, dear reader, to *Soil Biology in Tropical Ecosystems*. This handbook is divided into eight chapters, where we will show you the main concepts of soil science, soil biology, soil ecology, and some interactions among soil abiotic and biotic traits. Enjoy!

Coimbra, Portugal                                                                 Tancredo Souza
February 02, 2022

# About the Author

**Tancredo Souza** wrote this book while he was a visiting professor in the Department of Agriculture, Biodiversity, and Forests at the Federal University of Santa Catarina, Curitibanos, Brazil. His research focus on the effects of land use on soil organisms' diversity from tropical and subtropical ecosystems. His interests are in the plant-soil interactions, ecology of mycorrhizal fungi, and soil microbiology. He holds degrees in agronomy (BSc), soil and water management and conservation (MS), and soil sciences (PhD) from the Federal University of Paraíba, Brazil, and the University of Coimbra, Portugal.

# Chapter 1
# The Soil Ecosystem at the Tropics

**Abstract** In this chapter, the soil ecosystem is introduced as a multiphase system that (i) acts as habitat for a wide diversity of soil organisms and (ii) varies at both spatial and temporal scale. The soil formation varies according to a combination of geological factors and biological process (e.g., here included the influence of mankind) that result in an almost infinite variation in soil-forming factors. Basically, five forming factors are defined as the most important factors. These forming factors are parent material, climate, topography, time, and the activity of soil organisms (e.g., plant roots, insects, microorganisms, human influence, etc.). Considering the wide range of soil properties, i.e., physical, chemical, and biochemical variables, in this chapter we focused only on describing the most important and significant properties to soil organisms, such as soil organic carbon, soil pH, soil aggregation, and moisture. Finally, when considering the tremendous variety of soil types, the need for soil ecologist to recognize this variation must be considered to avoid stresses. Especially, if the student is considering both spatial and temporal variation into soil ecosystem. In view of this, it is important that soil ecologist must consider both soil biota and soil ecosystem characterization.

**Keywords** Soil as habitat · Soil ecosystem · Soil-forming factors · Soil physical and chemical properties · Soil spatial and temporal variation

**Questions Covered in the Chapter**
1. How can a soil ecologist define soil ecosystem?
2. How many soil types can we find at the tropics?
3. Define soil horizons and why do the organic horizons important for soil ecology?
4. What are the pedogenetic processes?
5. How the soil formation factors promote soil formation?
6. How soil properties affect soil organisms at the tropics?
7. Why is important to consider the ecosystem in studies considering soil biology or soil ecology?

© The Author(s), under exclusive license to Springer Nature Switzerland AG 2022    1
T. Souza, *Soil Biology in Tropical Ecosystems*, https://doi.org/10.1007/978-3-031-00949-5_1

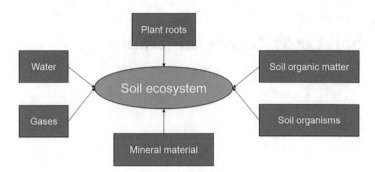

**Fig. 1.1** The multiphase nature of the soil ecosystem. (Adapted from Anyango et al. 2020; Bai et al. 2018; Balkenhol et al. 2018; and Zuo et al. 2019))

## 1.1   Introduction

Soil ecosystem is a multiphase system that covers the Earth's surface (Karima et al. 2020). It acts as (i) habitat for a wide range of soil organisms, (ii) nutrient-rich environment for plant growth, and (iii) a carbon and water reservoir (Hein et al. 2020; McGee et al. 2020). The soil ecosystem also consists of a pool of organic matter at various stages of decay that acts as medium in which most of soil organisms live (e.g., litter transformers, decomposers, prokaryotic transformers, and microregulators) (Souza and Freitas 2018). As a result, the soil ecosystem at the tropics is strongly influenced by the soil biota activity (Forstall-Sosa et al. 2020). The soil organisms use the soil as habitat and food resource, and in turn, they contribute to soil formation by affecting soil physical (e.g., ecosystems engineers improving soil porosity) and chemical properties (e.g., decomposers improving soil organic matter decomposition and nutrient cycling) (Melo et al. 2019). Indeed, the soil organism activity is one of the five forming factors: parent material, climate, topography, time, and the activity of soil organisms (Bonfatti et al. 2020). As the soil ecosystem is considered a multiphase system (Fig. 1.1), a soil ecologist needs to understand the physical and chemical properties of the soil matrix that modulate the abundance and community structure of soil organisms considering both the spatial and temporal variability (Hou et al. 2019). Thus, it is important to remind that soil organisms live into the soil ecosystem. Here, the soil ecologist will find a background on the soil formation and key soil properties that most affect soil organism community structure.

## 1.2   Soil Formation

A beginner in soil ecology must consider that there are factors affecting soil formation within the landscape. So, the task to understand how soil properties influence the soil biota activity and its community structure will become less stressful (da Silva et al. 2020). At the tropics, there is a tremendous variety of landforms (e.g., elevation, slope, rock exposure, and soil type) that were extremely affected by geological process and biological process that have occurred over millions of years and, more recently, respectively (Rahbek et al. 2019). The combination of these two processes resulted in an almost infinite variety of soil types around all the Earth's surface (Fujii et al. 2018), but following the *Keys to Soil Taxonomy* (Balota 2017; Doniger et al. 2020; Lammel et al. 2015; Lira et al. 2020; Machado et al. 2015; Manwaring et al. 2018; and Mauda et al. 2018) at the tropics, 11 soil types were described:

1. *Alfisols*: Soils with an argillic or kandic horizon that cover an area of 12.4% from the tropics. Alfisols present more than 35% base saturation at pH 8.2 to 1.8 m depth. They are formed over basic rocks or sediments. This soil type has an A–E–Bt–C horizon sequence and presents high native fertility.
2. *Andisols*: Soils developing in volcanic ash, pumice, cinders, and lava with andic properties. They present a high phosphorus retention and cover an area of 1.2% from the tropics. Usually, Andisols are very fertile but phosphorus.
3. *Aridisols*: Soils that are found exclusively in aridic moisture regimes (e.g., also called desert soils) that cover an area of 4.8% from the tropics. This soil type has an A–Bt–Bk or Bw–C horizon sequence.
4. *Entisols*: Soils of such slight development (e.g., an ochric (yellowish) epipedon) that cover an area of 15.6% from the tropics. This soil type has an A–C or A–C–R horizon sequences. Most Entisols are present on steep slopes, flood plains, or inert sandy parent material.
5. *Histosols*: Organic soils (> 12% of organic C to >60 cm) that cover an area of 0.8% from the tropics. They present low bulk density, low fertility, and wet (saturated >30 days each year or artificially drained) and subside when drained.
6. *Inceptisols*: Soils with only a cambic horizon that cover an area of 15.7% from the tropics. This soil type has an A–B–C horizon sequence. They are present in all soil moisture regimes but aridic. Inceptisols are most prevalent on sloping landscapes and relatively recent colluvial sediments.
7. *Mollisols*: Soils with >50% base saturation at pH 7 to a depth of 1.8 m, a mollic epipedon, and that cover an area of 0.9% from the tropics. This soil type has an A–Bw Bt or Bk–C horizon sequences. Histosols are very fertile and considered as the best agricultural lands.
8. *Oxisols*: Deep soils of sandy loam or finer texture that cover an area of 24.8% from the tropics. This soil type has strong granular structure. They can be found in all soil moisture regimes. The sand fraction is dominated by quartz, whereas the clay fraction consisted of mixtures of kaolinite, gibbsite, and aluminum and iron oxides. This soil type has an A–B–C horizon sequence. Many Oxisols are well-drained and have low fertility, but some are quite naturally fertile.
9. *Spodosols*: Soils with a spodic horizon (of iron and/or organic matter accumulation), usually developed on sandy materials, and that cover an area of 0.2% from the tropics. This soil type has an A–E–Bh or Bs–C horizon sequences.

10. *Ultisols*: Soils with an argillic or kandic horizon and low base saturation and that cover an area of 19.6% from the tropics. Most Ultisols are formed over acid igneous rocks or sediments derived from such rocks. This soil type has an A–E–Bt–C horizon sequence. Clay fraction consists mainly of kaolinite, gibbsite, and aluminum-interlayer clays. It has good but less desirable physical properties than Oxisols and relatively low native fertility.
11. *Vertisols*: Heavy, cracking clayey soils (e.g., >30% clay content in all horizons and > 50% of "2:1" minerals in clay fraction) that open and close periodically with changes in soil moisture. Vertisols cover an area of 3.9% from the tropics. This soil type has an A–Bss–C horizon sequence. Some have gilgai microrelief and slickensides.

Each soil type presents unique characteristics and horizon sequences. The soil horizons are layers that constitute the soil profile (Fig. 1.2). In general, five horizons (layers) are well defined by soil scientists, but for soil ecologist only the horizons that are close to the L layer are considered important once the most soil organisms live there and just into these layers the roots growth (e.g., improving rootability and rhizodeposition) and the nutrient cycling process occurs (Jat et al. 2021). These important layers to soil organisms are referred as organic horizons that were originated by the accumulation and decomposition of litter on soil surface (Ramírez et al. 2017; Cheng et al. 2020). Thus, organic horizons are strongly influenced by plant community, whereas the mineral horizons are strongly influenced by the weathering of the parental material.

First, we must consider the L layer as the most biologically active and functional layer of the soil profile (Cheng et al. 2021; Liu et al. 2021). Most of soil organisms from macrofauna group, i.e., litter transformers and predators, live in L layer (e.g., composed by leaf litter, dead wood) (Nanganoa et al. 2019). Here we also can find a small minority of decomposers (e.g., microbiota group) (Jo et al. 2020). According Myer and Forschler (2019), Ng et al. (2018), Paymaneh et al. (2019), and Purwanto

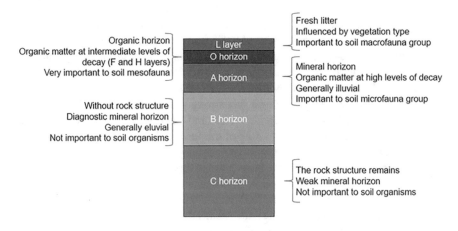

**Fig. 1.2**  A schematic view of a soil profile. (Adapted from Rahbek et al. (2019))

and Alam (2020), the L layer can affect soil organism abundance and diversity (measured by Shannon's index). The first one is affected by the L layer quantity, whereas the second one is affected by L layer quality. The organic horizon (F and H layers) that is considered the main habitat to soil organisms from mesofauna group (e.g., Acari, Collembola, Diplura, and Protura), and ecosystem engineers such as Blattodea (including Termitidae) and Hymenoptera (Coulis 2021). Some studies around the tropics have showed that organic soils usually present higher abundance of soil organisms, and the most important is the presence of high trophic levels of soil organisms into the soil food web (Marian et al. 2018; Pestana et al. 2020; Habel et al. 2021). Finally, the uppermost A horizon, which is composed by a mixture of mineral material and organic matter, is derived from L layer and O horizon. Here the microbiota group is the most important functional group acting primarily in the nutrient cycling, symbiosis, and microregulation processes (Certinia et al. 2021). Arbuscular mycorrhizal fungi plays an important role into the A horizon by improving both nutrient and water uptake by plant species with colonized roots (Basirat et al. 2019; Carrillo-Saucedo et al. 2018; Coyle et al. 2017; and de Novais et al. 2020).

Overall, tropical soils may vary across landscapes (da Silva et al. 2021a, 2021b). As described above, they can be classified into 11 groups at the tropics. This classification is based on quantitatively defined diagnostic horizons and attributes that are known to exist in nature (Kyebogola et al. 2020). So, first we need to know the soil ecosystem describing their unique characteristics and understanding how these characteristics can modulate soil organism community structure (Bai et al. 2021). Soil classifications provide to us a clear and concise identification grouping soils according their soil pedogenetic processes (Swift et al. 1979; and Tunlid and White 1992). Basically, according to Hildebrandt-Radke et al. (2017), we can define five main pedogenetic processes:

1. Gleying: It occurs in hydromorphic soils that are characterized by waterlogging for long period. This pedogenetic process occurs by biological means (e.g., microbial reduction of $Fe^{2+}$ to ferrous$^{3+}$ under anaerobic conditions), and it is characterized by the presence of a gleyed horizon.
2. Leaching: It occurs in well-drained soils with high content of mobile nutrients. This mobility depends on nutrient solubility in water, their relationship with soil pH, and the rate of water percolation. This pedogenetic process occurs by physical and chemical means (e.g., downward movement of nutrients in acid soil solution from one soil horizon to another).
3. Lessivage: It occurs under conditions that favor deflocculation of clay minerals. Here, clay types are translocated down to the soil profile creating a subsurface horizon that is referred as $B_t$ or argillic horizon. This pedogenetic process occurs by chemical (e.g., clay mineral deflocculation in the presence of $Na^+$ or hydrophilic organic films in acid soils and hence move in water) and physical (e.g., when clay minerals are destabilized by physical impact, such as raindrops) means.

4. Podzolization: It occurs under high leaching of Al and Fe from upper mineral horizon to deep mineral horizons into soil profile. Here, the leaching of these elements is enhanced by soluble organometal complexes, also referred as chelates. This pedogenetic process occurs mainly by chemical means (e.g., removal of $Al^{3+}$ or $Fe^{2+}$ promoted by organic acids and polyphenols, thus creating an eluvial horizon ($E_a$) and a spodic horizon ($B_s$).
5. Weathering: It occurs by the action of a wide range of combined forces that break up rock in small particles. These particles are classified as the parent material of soil. This pedogenetic process occurs by physical (e.g., by the action of wind, water, and temperature that break down rock into small particles), chemical (e.g., by the decomposition of minerals), and biological means (e.g., by the biological activity).

For example, tropical soils formed by podzolization (Usually Spodosols) present a deep soil profile, low decomposition rates, low fertility, and a very acid surface. Typically, this surface is an O horizon. The absence of earthworms, the high number of Acari and Collembola, and a microbial biomass mostly dominated by fungi are the main characteristic of soil organism community structure into these soils (Woese 1987; Kamczyc et al. 2017; Fujii et al. 2018; Stoops et al. 2020)

## 1.3   Soil-Forming Factors

The nature of soils and their development was influenced by soil formation factors (Bardgett 2005; Devaney et al. 2020; and Dial et al. 2006). They are climate, organisms (vegetation, mankind, and soil biota), parent material, topography, and time (Kowalchuck et al. 1997; Raiesi & Salek-Gilani 2020; Rasmussen et al. 2019; and Real-Santillán et al. 2019). The 11 tropic soil types vary accordingly to the intensity that the forming factors interacted with themselves. This section summarizes some of the important aspects related to each soil-forming factors.

### 1.3.1   Climate

The climate affects soil formation due to temperature and precipitation. For temperature, between the Tropics of Cancer and of Capricorn, there are small differences in temperature during the years (e.g., mean annual temperatures are normally above 25 °C) (Wood et al. 2019). The soil temperature regime in the tropics is described as hyperthermic (approximately 87% of the tropics area), but there are some places where the soil temperatures is less than 6 °C, thus being described as isohyperthermic (Abubakar et al. 2019; Currylow et al. 2021). However, for precipitation there are big differences in terms of the rainfall during the year and, consequently, in the soil moisture. Basically, there are three tropical climate types that are classified and distinguished by the annual precipitation level:

1. Tropical rainforest climate (Af): It presents an annual precipitation with more than 3500 mm, well distributed through the year. The Af type covers approximately 24% of the tropics, and it also presents high temperatures. The soil moisture regime here is described as *udic*.
2. Tropical monsoon climate (Am): It presents an annual precipitation near to 3000 mm with 90% of the rainfall concentrated within the summer and the other 10% within the winter (e.g., commonly a short-drought season occurs in this season). The Am type covers approximately 49% of the tropics. The soil moisture regime here is described as *ustic*.
3. Tropical wet and dry or just Savanna climate (Aw or As): It presents an annual precipitation ranging from 700 to 1000 mm during the year. The areas with these climate types present the driest month (e.g., with less than 60 mm of rainfall) ranging from November to March. This climate type covers approximately 27% of the tropics. The soil moisture regime here is described as *torric*.

Both precipitation and temperature strongly affect weathering, decomposition of organic matter, soil chemical reactions, plant growth, and soil biota activity and abundance (Zhou et al. 2018; Sato et al. 2019; Heydari et al. 2020; and Silva & Siqueira 2020). The effects of these two variables also vary according to altitude (e.g., being higher at high altitude and latitudes), spatial scale (e.g., being especially higher at vegetated soils), and other environmental factors (Xu et al. 2018; Liu et al. 2019).

## 1.3.2   Parent Material

The soil ecosystem is derived from the weathering of the rock and sedimentary deposits, i.e., consolidated rock in situ and unconsolidated rock, respectively (Raj 2018; Malone and Searle 2020). There are three properties of the parent material that affect the soil formation. They are degree of consolidation (e.g., fissured rocks will offer a large surface area for weathering than impermeable rocks, thus the weathering will occur more extensively throughout the material of the fissured rocks), the particle size (e.g., determining the soil texture based on the proportions of clay, sand, and silt), and parent material composition (e.g., defining the nature of the soil and its potential uses) (Loba et al. 2020; Peng et al. 2020a; Santos et al. 2021).

## 1.3.3   Topography

The topography affects soil formation due to effects on soil drainage (e.g., based on the position of a soil on a slope) and erosion (e.g., greatly influenced by the movement of soil particles downslope by the action of water and wind) (Conforti et al. 2020; Sadiq et al. 2021). Also, the topography has an indirect influence on climate (e.g., for every 1000 m increase in altitude, the temperature falls in 6 °C) that in turn

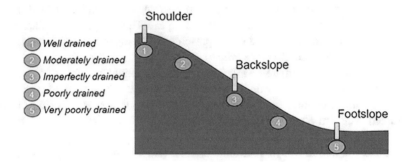

**Fig. 1.3** Changes in soil hydrological patterns according the slope position. (Adapted from Molina et al. (2019), Peng et al. (2020a, b), Sadiq et al. (2021))

greatly affects the chemical reactions and biological activity into soil profile (Molina et al. 2019; Peng et al. 2020b). Finally, the topography determines the soil hydrological variation within a landscape (Fig. 1.3).

### 1.3.4 Organisms

The soil organisms, the vegetation, and mankind constitute the main soil-forming factor (González-Gusmán et al. 2017; Rodriguez-Caballero et al. 2018; Wietrzyk-Pelka et al. 2020). The combination of these three components affects soil formation and profile development (Mikutta et al. 2019). Within soil organism group, ecosystem engineers, litter transformers, decomposers, and prokaryotic transformers are particularly important to transform and incorporate litter, to improve soil organic matter compartment, and to promote nutrient cycling. For example, litter transformers break litter down physically (e.g., by their mandibles activity), while decomposers break the organic residues down chemically (e.g., by enzymatic activity) (Kamau et al. 2017; Santos-Heredia et al. 2018; Sahu et al. 2019; and Rodríguez et al. 2019). So, the activity of soil organisms contributes to (i) release plant nutrients and (ii) produce organic products that affect positively soil aggregate stability (Escudero-Martinez and Bulgarelli 2019; Plaas et al. 2019).

The vegetation also induces changes into soil profile through its root activity that creates a unique environment into soil profile called rhizosphere (Ma et al. 2018). Commonly, the rhizosphere presents more C- and N-rich compounds than a non-rhizospheric soil, thus attracting a wide range of soil organisms. This process is called rhizodeposition (Zhang et al. 2020; da Silva et al. 2021a, b). The plant roots can penetrate and crack open rocks. Also, the vegetation plays an important role covering the soil surface. So, the vegetation acts protecting the soil surface against the erosion (Flores et al. 2020). The vegetation is the main source for the input of L layer (Nakayama et al. 2019).

Finally, the human influence promotes even more significant impacts on soil formation through land use and management (Baude et al. 2019). Most of human activities promote negative impacts on ecosystem, such as biological invasion promoted by
the destruction of natural vegetation, overlogging, fire, natural disasters promoted by
climate change, soil C losses, illegal timber extraction, agricultural conversion, and
severe soil degradation. All these activities negatively change both the quality and
quantity of litter, soil organic matter compartment, and nutrient cycling (Deng
et al. 2018).

### 1.3.5  Time

Time is the unique soil-forming factor that acts different from the other four soil-
forming factors and distinguishes tropical soils from the cool temperate climates. Its
influence is related to the duration of environmental factors on the Earth's surface
(Alekseev et al. 2019). We must remember that the time acts forming soils since the
Earth's formation (Yu and Hunt 2018). Here, some laws can be defined considering
only the effect of time on soil formation over time:

1. Soils become more weathered.
2. Soil profile becomes progressively deeper and differentiated.
3. Within landscape, footslope soils become more fertile than shoulder soils.
4. Clay minerals become leached down profile.
5. Subsurface argillic horizons become more frequent.
6. Increases in soil organic matter and N content are present.
7. The availability of P into soil profile decreases significantly by its loss and fixation in mineral forms.

## 1.4  Soil Properties

Soil properties are determined by the interaction of soil-forming factors over time
(Lucas et al. 2019), which in turn affects soil organism community structure (Guo
et al. 2018). This community will provide services into soil ecosystem such as biological control, decomposition, nutrient cycling, and symbiosis (Zhang et al. 2021).
Thus, variation in soil chemical and physical properties may have a great influence
on soil organism groups (Dai et al. 2021). This section will describe the most important and significant properties to soil organisms, such as soil organic carbon, soil
pH, soil aggregation, and moisture. Accordingly, to Souza and Freitas (2018), all
these variables modulate soil organism activity, abundance, diversity, and behavior
into soil profile.

### 1.4.1   Soil Organic Carbon

The soil organic carbon content of tropical soils varies significantly in terms of its quantity, soil type, soil C pools, and above- and belowground vegetation activity (Sun et al. 2020a, b; Rajapaksha et al. 2020). This last one being considered as the most important once, it determines the quantity of C inputs through litter deposition (Maas et al. 2021; Oyedeji et al. 2021; Xu et al. 2021). Other factors, such as climate, soil drainage, and soil organism activity (e.g., litter transformers and decomposers), can influence soil organic carbon into soil profile (Souza et al. 2016; Frouz 2018; Sousa et al. 2018; Souza 2018; Yang et al. 2018; Gross and Harrison 2019; and Suleiman et al. 2019). Tropical soils with high contents of soil organic carbon tend to have high structural stability (e.g., here providing a high stable habitat to soil organisms), high water content (e.g., by its property to retain water), and an active soil biota (e.g., once soil organic carbon acts as a primary source of C to soil organisms) (Kwiatkowska-Malina 2018; Campos et al. 2020).

In tropical soils, carbon cycle starts with litter deposition that its quality and quantity is modulated by plant community structure and net primary production (Queiroz et al. 2019; Luo et al. 2021). Next, the litter accumulated on soil surface is broken down physically by their mandible activity (Magcale-Macandog et al. 2018; Cole et al. 2019). Also, litter transformers help to incorporate this material into soil profile, thus creating F and H layers (Chervot et al. 2017; Potatov et al. 2017; Sofo et al. 2020). In soil profile, these layers will still be physically transformed to C-rich particles (e.g., cellulose, amino sugars, nucleic acid, lignin, etc.) that will be more accessible to microbial chemical attack (Pastorelli et al. 2021; Zhang et al. 2019). These particles constitute the soil C pool into soil organic matter. Decomposers will convert C-rich particles into simple organic molecules, thus producing $CO_2$ from heterotrophic microbial activity (Girkin et al. 2018; Tabrizi et al. 2022). The overall soil $CO_2$ evolves into the atmosphere, and the latter will be incorporated into vegetation biomass through photosynthesis process (Fig. 1.4).

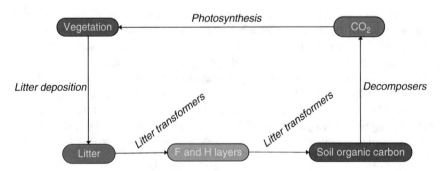

**Fig. 1.4** Carbon cycle into tropical soils and the role of soil organisms. (Adapted from Souza and Freitas (2018))

## 1.4.2 Soil pH

Soil reaction (pH) determines soil nutrient availability (e.g., $K^+$, $Ca^{+2}$, $Mg^{2+}$, P, and $S$-$SO_4^{2-}$) into soil profile, which in turn directly affects soil organisms and plant growth (Hou et al. 2018; Hong et al. 2018; Li et al. 2019). It also varies from a horizon to another. At the tropics, we can find a wide variation of soil pH by each soil type (Negasa et al. 2017; Hou et al. 2019). We can find acid soils associated with tropical rainforest climate (Muchane et al. 2020; Maru et al. 2020). On the other hand, at the Savanna climate, there are saline, saline-alkali, and non-saline-alkali soils (e.g., by salinization, sodium predominance, and alkalinization processes) (Meira-Neto et al. 2017; Bayoumi et al. 2021; Oliveira et al. 2021).

In general, acid soils come from four main sources, such as carbonic acid, microbial activity, H+ extrusion, pollution, rootability, and soil organic matter decomposition (Zou et al. 2018; Han et al. 2020; Barbosa et al. 2021). Saline soils come from the natural excess of sodium salts into soil profile without damage in the soil colloids (Saifullah et al. 2018; Zhao et al. 2018). Saline-alkali soils come when sodium starts accumulating and predominating over the other cations (e.g., $Ca^{2+}$, $Mg^{2+}$, and $K^+$) into soil profile over a long period promoting clay flocculation in excess (El hasini et al. 2019; Mahmoud et al. 2019). Finally, non-saline-alkali soils come when only Na remains into soil colloids (e.g., Na-Clay) (Litalien and Zeeb 2020; Wei et al. 2020).

## 1.4.3 Soil Aggregation

Soil aggregation determines the physical structure of soil as a habitat for soil organisms (Domínguez et al. 2018; Zheng et al. 2018). This variable estimates the pore distribution (e.g., good soil aggregation prevents anaerobic conditions into soil profile) and aggregates diameter into soil profile (e.g., good distribution of aggregates promotes soil organism movement) (Velasquez and Lavelle 2019; Franco et al. 2020). It affects soil water distribution and the way that soil organisms live, thus affecting the soil food web by promoting trophic interactions (Lehmann et al. 2017; Chen et al. 2019). Also, it is strongly affected by soil organism activity in three main ways:

1. Soil glomalin: It is a glycoprotein produced by arbuscular mycorrhizal fungi that act binding mineral particles together, thus improving soil aggregation.
2. Mucigel: It is a highly hydrated polysaccharide produced by the outermost cells of the root cap that act creating the rhizosphere (e.g., by creating a layer of microbiota-rich around the mucigel). It improves root penetration into soil and soil water movement, thus improving soil aggregation.
3. Microbial exopolysaccharide (EPS): It is a carbohydrate polymer with high molecular weight produced by microbiota activity. EPS acts as carbon and energy reserves for soil microbiota. Its viscous characteristic improves soil moisture retention, soil biofilm, and soil aggregation, thus promoting soil strengthening.

### 1.4.4   Soil Moisture (Soil Water)

Water is the main factor that creates favorable conditions into the soil to render it as a habitable environment for soil organisms (Roy et al. 2018; Olatunji et al. 2019). Into the soil pores, the water acts mainly by dissolving nutrients and creating biofilms where the soil microbiota lives and moves around in (Chen et al. 2020; Singh et al. 2021; Wale and Yesuf 2021). Into this context, we must consider that soil organisms are extremely sensitive to desiccation; thus, soil water becomes a key factor to render the soil as a favorable habitat to soil organisms. So, it is common to refer soil water as soil solution (Kabala et al. 2017; Smagin et al. 2018). Also, this soil solution supplies plant species with water and nutrients (Gebremikael et al. 2016; Pasley et al. 2019; Thakur et al. 2021)

## 1.5   Conclusion

In tropical ecosystem, there are 11 soil types that vary according to their physical and chemical characteristics. Understating the soil formation or pedogenetic process, the five soil formation factors, and the properties which strongly affect soil organisms at tropical ecosystem is a hard task and a great challenge for soil ecologist. We must consider that the soil ecosystem acts both as habitat and food resource for a complex soil food web that interacts constantly at trophic levels. Nevertheless, we must try to understand soil profile (e.g., especially L layer and organic horizon) where the soil organisms live and interact with each other. Finally, when considering the tremendous variety of soil types, the need for soil ecologist to recognize this variation must be considered to avoid stresses. Studies considering soil organism community structure accompanied for a detailed soil characterization are welcome. Especially, if the student is considering both spatial and temporal variation into soil ecosystem. In view of this, it is important that soil ecologist must consider both soil biota (e.g., identifying taxonomic levels, functional groups, and their behavior) and soil ecosystem characterization (e.g., by providing information about physical and chemical properties).

## References

Abubakar MA, Zega ZM, Jayeoba OJ (2019) Major characteristics and classification of soils in Duduguru, obi LGA of Nasarawa State, Nigeria. Agric Res J 56:417–423. https://doi.org/10.5958/2395-146X.2019.00067.X

Alekseev AO, Kalinin PI, Alekseeva TV (2019) Soil indicators of paleoenvironmental conditions in the south of the east European plain in the quaternary time. Eurasian Soil Sci 52:349–358. https://doi.org/10.1134/S1064229319040021

Anyango JJ, Bautze D, Fiaboe KKM, Lagat ZO, Muriuki AW, Stöckli S, Riedel J, Onyambu GK, Musyoka MW, Karanja EN, Adamtey N (2020) The impact of conventional and organic farming on soil biodiversity conservation: a case study on termites in the long-term farming systems comparison trials in Kenya. BMC Ecol. https://doi.org/10.1186/s12898-020-00282-x

Bai P, Liu Q, Li X, Liu Y, Zhang L (2018) Response of the wheat rhizosphere soil nematode community in wheat/walnut intercropping system in Xinjiang, Northwest China. Appl Entomol Zool. https://doi.org/10.1007/s13355-018-0557-9

Bai X, Wang J, Dong H, Chen J, Ge Y (2021) Relative importance of soil properties and heavy metals/metalloids to modulate microbial community and activity at a smelting site. J Soil Sediments 21:1–12. https://doi.org/10.1007/s11368-020-02743-8

Balkenhol B, Haase H, Gebauer P, Lehmitz R (2018) Steeplebushes conquer the countryside: influence of invasive plant species on spider communities (Araneae) in former wet meadows. Biodivers Conserv 27:2257–2274. https://doi.org/10.1007/s10531-018-1536-8

Balota EL (2017) Manejo e qualidade biológica do solo. Editora Mecenas, Londrina

Barbosa LS, Souza TAF, Lucena EO, da Silva LJR, Laurindo LK, Nascimento GS, Santos D (2021) Arbuscular mycorrhizal fungi diversity and transpiratory rate in long-term field cover crop systems from tropical ecosystem, northeastern Brazil. Symbiosis. https://doi.org/10.1007/s13199-021-00805-0

Bardgett RD (2005) The biology of soil: a community and ecosystem approach. Oxford University Press, New York

Basirat M, Mousavi SM, Abbaszadeh S, Ebrahimi M, Zarebanadkouki M (2019) The rhizosheath: a potential root trait helping plants to tolerate drought stress. Plant Soil 445:565–575. https://doi.org/10.1007/s11104-019-04334-0

Baude M, Meyer BC, Schindewolf M (2019) Land use change in an agricultural landscape causing degradation of soil based ecosystem services. Sci Total Environ 659:1526–1536. https://doi.org/10.1016/j.scitotenv.2018.12.455

Bayoumi Y, Abd-Alkarim E, El-Ramady H, El-Aidy F, Hamed E-S, Taha N, Prohens J, Rakha M (2021) Grafting improves fruit yield of cucumber plants grown under combined heat and soil salinity stresses. Horticulturae 7(3):61. https://doi.org/10.3390/horticulturae7030061

Bonfatti BR, Demattê JAM, Marques KPP, Poppiel RR, Rizzo R, Mendes WS, Silvero NEQ, Safanelli JL (2020) Digital mapping of soil parent material in a heterogeneous tropical area. Geomorphology 367:e107305. https://doi.org/10.1016/j.geomorph.2020.107305

Campos AC, Suárez GM, Laborde J (2020) Analyzing vegetation cover-induced organic matter mineralization dynamics in sandy soils from tropical dry coastal ecosystems. Catena 185:104264. https://doi.org/10.1016/j.catena.2019.104264

Carrillo-Saucedo SM, Gavito ME, Siddique I (2018) Arbuscular mycorrhizal fungal spore communities of a tropical dry forest ecosystem show resilience to land-use change. Fungal Ecol. https://doi.org/10.1016/j.funeco.2017.11.006

Certinia G, Moya D, Lucas-Borja ME, Mastrolonardo G (2021) The impact of fire on soil-dwelling biota: a review. For Ecol Manag. https://doi.org/10.1016/j.foreco.2021.118989

Chen Y, Cao J, He X, Liu T, Shao Y, Zhang C, Zhou Q, Li F, Mao P, Tao L, Liu Z, Lin Y, Zhou L, Zhang W, Fu S (2020) Plant leaf litter plays a more important role than roots in maintaining earthworm communities in subtropical plantations. Soil Biol Biochem 144:107777. https://doi.org/10.1016/j.soilbio.2020.107777

Chen Y, Hu N, Zhang Q, Lou Y, Li Z, Tang Z, Kuzyakov Y, Wang Y (2019) Impacts of green manure amendment on detritus micro-food web in a double-rice cropping system. Appl Soil Ecol 138:32–36. https://doi.org/10.1016/j.apsoil.2019.02.013

Cheng J, Ma W, Hao B, Liu X, Li FY (2021) Divergent responses of nematodes in plant litter versus in top soil layer to nitrogen addition in a semi-arid grassland. Appl Soil Ecol. https://doi.org/10.1016/j.apsoil.2020.103719

Cheng Y, Wang J, Wang J, Wang S, Chang SX, Cai Z, Zhang J, Niu S, Hu S (2020) Nitrogen deposition differentially affects soil gross nitrogen transformations in organic and mineral horizons. Earth Sci Rev 201:103033. https://doi.org/10.1016/j.earscirev.2019.103033

Chervot O, Komarov A, Shaw C, Bykhovets S, Frolov P, Shanin V, Grabarnik P, Priputina I, Zubkova E, Shashkov (2017) Romul_Hum—a model of soil organic matter formation coupling with soil biota activity. II. Parameterisation of the soil food web biota activity. Ecol Model 345: 125-139. https://doi.org/10.1016/j.ecolmodel.2016.10.024

Cole RJ, Selmants P, Khan S, Chazdon R (2019) Litter dynamics recover faster than arthropod biodiversity during tropical forest succession. Biotropica 52(1):22–33. https://doi.org/10.1111/btp.12740

Conforti M, Longobucco T, Scarciglia F, Niceforo G, Matteucci G, Buttafuoco G (2020) Interplay between soil formation and geomorphic processes along a soil catena in a Mediterranean mountain landscape: an integrated pedological and geophysical approach. Environ Earth Sci 79:59. https://doi.org/10.1007/s12665-019-8802-2

Coulis M (2021) Abundance, biomass and community composition of soil saprophagous macrofauna in conventional and organic sugarcane fields. Appl Soil Ecol 164:103923. https://doi.org/10.1016/j.apsoil.2021.103923

Coyle DR, Nagendra UJ, Taylor MK, Campbell JH, Cunard CE, Joslin AH, Mundepi A, Phillips CA, Callaham MA Jr (2017) Soil fauna responses to natural disturbances, invasive species, and global climate change: current state of the science and a call to action. Soil Biol Biochem. https://doi.org/10.1016/j.soilbio.2017.03.008

Currylow AF, Collier MAM, Hanslowe EB, Falk BG, Cade BS, Moy SE, Grajal-Puche A, Ridgley FN, Reed RN, Adams AAY (2021) Thermal stability of an adaptable, invasive ectotherm: Argentine giant tegus in the Greater Everglades ecosystem, USA. Ecosphere 12(9):e03579. https://doi.org/10.1002/ecs2.3579

da Silva JVCLS, Hirschfeld MNC, Cares JE, Esteves AM (2020) Land use, soil properties and climate variables influence the nematode communities in the Caatinga dry forest. Appl Soil Ecol. https://doi.org/10.1016/j.apsoil.2019.103474

da Silva SIA, Souza TAF, Lucena EO, da Silva LJR, Laurindo LK, Nascimento GS, Santos D (2021a) High phosphorus availability promotes the diversity of arbuscular mycorrhizal spores' community in different tropical crop systems. Biologia. https://doi.org/10.1007/s11756-021-00874-y

da Silva TGF, de Queiroz MG, Zolnier S, de Souza LSB, de Souza CAA, de Moura MSB, de Araújo GGL, Steidle Neto AJ, dos Santos TS, de Melo AL, Cruz Neto JF, da Silva MJ, Alves HKMN (2021b) Soil properties and microclimate of two predominant landscapes in the Brazilian semi-arid region: comparison between a seasonally dry tropical forest and a deforested area. Soil Tillage Res 207:104852. https://doi.org/10.1016/j.still.2020.104852

Dai Z, Xiong X, Zhu H, Xu H, Leng P, Li J, Tang C, Xu J (2021) Association of biochar properties with changes in soil bacterial, fungal and fauna communities and nutrient cycling processes. Biochar 3:239–254. https://doi.org/10.1007/s42773-021-00099-x

de Novais CB, Sbrana C, Jesus EC, Rouws LFM, Giovannetti M, Avio L, Siqueira JO, Saggin Júnior OJ, da Silva EMR, de Faria SM (2020) Mycorrhizal networks facilitate the colonization of legume roots by a symbiotic nitrogen-fixing bacterium. Mycorrhiza. https://doi.org/10.1007/s00572-020-00948-w

Deng L, Wang K, Zhu G, Liu G, Chen L, Shangguan Z (2018) Changes of soil carbon in five land use stages following 10 years of vegetation succession on the Loess Plateau, China. Catena 171:185–192. https://doi.org/10.1016/j.catena.2018.07.014

Devaney JL, Marone D, McElwain JC (2020) Impact of soil salinity on mangrove restoration in a semi-arid region: a case study from the Saloum Delta, Senegal. Restor Ecol. https://doi.org/10.1111/rec.13186

Dial RJ, Ellwood MDF, Turner EC, Foster WA (2006) Arthropod abundance, canopy structure, and m microclimate in a Bornean lowland tropical rain forest. Biotropica 38(5):643–652

Domínguez A, Jiménez JJ, Ortiz CE, Bedano JC (2018) Soil macrofauna diversity as a key element for building sustainable agriculture in Argentine Pampas. Acta Oecol 92:102–116. https://doi.org/10.1016/j.actao.2018.08.012

Doniger T, Adams JM, Marais E, Maggs-Kölling G (2020) The "fertile island effect" of *Welwitschia* plants on soil microbiota is influenced by plant gender. FEMS Microbiol. https://doi.org/10.1093/femsec/fiaa186

El Hasini S, Halima OI, El Azzouzi M, Douiak A, Azim K, Zouahri A (2019) Organic and inorganic remediation of soils affected by salinity in the Sebkha of Sed El Mesjoune – Marrakech (Morocco). Soil Tillage Res 193:153–160. https://doi.org/10.1016/j.still.2019.06.003

Escudero-Martinez C, Bulgarelli D (2019) Tracing the evolutionary routes of plant–microbiota interactions. Curr Opin Microbiol 49:34–40. https://doi.org/10.1016/j.mib.2019.09.013

Flores BM, Staal A, Jakovac CC, Hitora M, Holmgren M, Oliveira RS (2020) Soil erosion as a resilience drain in disturbed tropical forests. Plant Soil 450:11–25. https://doi.org/10.1007/s11104-019-04097-8

Forstall-Sosa KS, Souza TAF, Lucena EO, Silva SAI, Ferreira JTA, Silva TN, Santos D, Niemeyer JC (2020) Soil macroarthropod community and soil biological quality index in a green manure farming system of the Brazilian semi-arid. Biologia. In press. https://doi.org/10.2478/s11756-020-00602-y

Franco ALC, Cherubin MR, Cerri CEP, SIx J, Wall DH, Cerri CC (2020) Linking soil engineers, structural stability, and organic matter allocation to unravel soil carbon responses to land-use change. Soil Biol Biochem 150: 107998. https://doi.org/10.1016/j.soilbio.2020.107998

Frouz J (2018) Effects of soil macro- and mesofauna on litter decomposition and soil organic matter stabilization. Goederma 332:161–172. https://doi.org/10.1016/j.geoderma.2017.08.039

Fujii K, Shibata M, Katajima K, Ichie T, Katayama K, Turner BL (2018) Plant–soil interactions maintain biodiversity and functions of tropical forest ecosystems. Ecol Res 33:149–160. https://doi.org/10.1007/s11284-017-1511-y

Gebremikael MT, Steel H, Buchan D, Bert W, Neve S (2016) Nematodes enhance plant growth and nutrient uptake under C and N-rich conditions. Sci Rep. https://doi.org/10.1038/srep32862

Girkin NT, Turner BL, Ostle N, Craigon J, Siögerten S (2018) Root exudate analogues accelerate $CO_2$ and $CH_4$ production in tropical peat. Soil Biol Biochem 117:48–55. https://doi.org/10.1016/j.soilbio.2017.11.008

González-Gusmán A, Oliva M, Souza-Júnior VS, Peréz-Alberti A, Ruiz-Fernández J, Otero XL (2017) Biota and geomorphic processes as key environmental factors controlling soil formation at Elephant Point, Maritime Antarctica. Geoderma 300:32–43. https://doi.org/10.1016/j.geoderma.2017.04.001

Gross CD, Harrison RB (2019) The case for digging deeper: Soil organic carbon storage, dynamics, and controls in our changing world. Soil Syst 3(2):28. https://doi.org/10.3390/soilsystems3020028

Guo Y, Chen X, Wu Y, Zhang L, Cheng J, Wei G, Lin Y (2018) Natural revegetation of a semi-arid habitat alters taxonomic and functional diversity of soil microbial communities. Sci Total Environ 635:598–606. https://doi.org/10.1016/j.scitotenv.2018.04.171

Habel JC, Koc E, Gerstmeier R, Gruppe A, Seibold S, Ulrich W (2021) Insect diversity across an afro-tropical forest biodiversity hotspot. J Insect Conserv 25:221–228. https://doi.org/10.1007/s10841-021-00293-z

Han M, Sun L, Gan D, Fu L, Zhu B (2020) Root functional traits are key determinants of the rhizosphere effect on soil organic matter decomposition across 14 temperate hardwood species. Soil Biol Biochem 151:108019. https://doi.org/10.1016/j.soilbio.2020.108019

Hein CJ, Usman M, Eglinton TI, Haghipour N, Galy VV (2020) Millennial-scale hydroclimate control of tropical soil carbon storage. Nature 581:63–66. https://doi.org/10.1038/s41586-020-2233-9

Heydari M, Eslaminejad P, Kakhki FV, Mirab-Balou M, Omidipour R, Prévosto B, Kooch Y, Lucas-Borja MEL (2020) Soil quality and mesofauna diversity relationship are modulated by woody species and seasonality in semiarid oak forest. For Ecol Manag. https://doi.org/10.1016/j.foreco.2020.118332

Hildebrandt-Radke I, Makarowicz P, Matviisgyna ZN, Parkhomenko A, Lysenko SD, Kochkin IT (2017) Late Neolithic and Middle Bronze Age barrows in Bukivna, Western Ukraine as a source to understand soil evolution and its environmental significance. J Archaeol Sci Rep 29:101972. https://doi.org/10.1016/j.jasrep.2019.101972

Hong S, Piao S, Chen A, Liu Y, Liu L, Peng S, Sardans J, Sun Y, Peñuelas J, Zeng H (2018) Afforestation neutralizes soil pH. Nat Commun 9:520. https://doi.org/10.1038/s41467-018-02970-1

Hou E, Wen D, Kuang Y, Cong J, Chen C, He X, Heenan M, Lu H, Zhang Y (2018) Soil pH predominantly controls the forms of organic phosphorus in topsoils under natural broadleaved forests along a 2500 km latitudinal gradient. Geoderma 315:65–74. https://doi.org/10.1016/j.geoderma.2017.11.041

Hou X, Han H, Tigabu M, Cai L, Meng F, Liu A, Ma X (2019) Changes in soil physicochemical properties following vegetation restoration mediate bacterial community composition and diversity in Changting, China. Ecol Eng 138:171–179. https://doi.org/10.1016/j.ecoleng.2019.07.031

Jat HS, Datta A, Choudhary M, Sharma PC, Dixit B, Jat ML (2021) Soil enzymes activity: effect of climate smart agriculture on rhizosphere and bulk soil under cereal based systems of northwest India. Eur J Soil Biol 103:103292. https://doi.org/10.1016/j.ejsobi.2021.103292

Jo I, Fridley JD, Frank DA (2020) Rapid leaf litter decomposition of deciduous understory shrubs and lianas mediated by mesofauna. Plant Ecol 221:63–68. https://doi.org/10.1007/s11258-019-00992-3

Kabala C, Karczewska A, Gałka B, Cuske M, Sowiński J (2017) Seasonal dynamics of nitrate and ammonium ion concentrations in soil solutions collected using MacroRhizon suction cups. Environ Monit Assess 189:304. https://doi.org/10.1007/s10661-017-6022-3

Kamau S, Barrios E, Karanja NK, Ayuke FO, Lehmann J (2017) Soil macrofauna abundance under dominant tree species increases along a soil degradation gradient. Appl Soil Ecol. https://doi.org/10.1016/j.soilbio.2017.04.016

Kamczyc J, Urbanowski C, Pers-Kamczyc E (2017) Mite communities (Acari: Mesostigmata) in young and mature coniferous forests after surface wildfire. Exp Appl Acarol 72:145–160. https://doi.org/10.1007/s10493-017-0148-4

Karima MR, Sultana F, Saimuna MSR, Mukul AS, Arfin-Khana MAS (2020) Plant diversity and local rainfall regime mediate soil ecosystem functions in tropical forests of north-east Bangladesh. Environ Adv. https://doi.org/10.1016/j.envadv.2020.100022

Kowalchuck GA, Stephen JR, de Boer W, Prosser JI, Embley TM, Woldendorp JW (1997) Analysis of ammonia-oxidising bacteria of the β-subdivision of the class *Proteobacteria* in coastal sand dunes by denaturing gradient gel electrophoresis and sequencing of PCR-amplified 16S ribosomal DNA fragments. Appl Environ Microbiol 63:1489–1497

Kwiatkowska-Malina J (2018) Qualitative and quantitative soil organic matter estimation for sustainable soil management. J Soils Sediments 18:2801–2812. https://doi.org/10.1007/s11368-017-1891-1

Kyebogola S, Burras LC, Miller BA, Semalulu O, Yost RS, Tenywa MM, Lenssen AW, Kyomuhendo P, Smith C, Luswata CK, Majaliwa MJG, Goettsch L, Colfer CJP, Mazur RE (2020) Comparing Uganda's indigenous soil classification system with world Reference Base and soil taxonomy. Geoderma Reg. https://doi.org/10.1016/j.geodrs.2020.e00296

Lammel DR, Cruz LM, Mescolotti D, Stürmer SL, Cardoso EJBN (2015) Woody Mimosa species are nodulated by Burkholderia in ombrophilous forest soils and their symbioses are enhanced by arbuscular mycorrhizal fungi (AMF). Plant Soil. https://doi.org/10.1007/s11104-015-2470-0

Lehmann A, Zheng W, Rillig MC (2017) Soil biota contributions to soil aggregation. Nat Ecol Evol 1(12):1828–1835. https://doi.org/10.1038/s41559-017-0344-y

Li Y, Cui S, Chang SX, Zhang Q (2019) Liming effects on soil pH and crop yield depend on lime material type, application method and rate, and crop species: a global meta-analysis. J Soils Sediments 19:1393–1406. https://doi.org/10.1007/s11368-018-2120-2

Lira AFA, Vieira AGT, Olveira RF (2020) Seasonal influence on foraging activity of scorpion species (Arachnida: Scorpiones) in a seasonal tropical dry forest remnant in Brazil. Stud Neotropical Fauna Environ. https://doi.org/10.1080/01650521.2020.1724497

Litalien A, Zeeb B (2020) Curing the earth: a review of anthropogenic soil salinization and plant-based strategies for sustainable mitigation. Sci Total Environ 698:134235. https://doi.org/10.1016/j.scitotenv.2019.134235

Liu L, Wang Z, Wang Y, Zhang Y, Shen J, Qin D, Li S (2019) Trade-off analyses of multiple mountain ecosystem services along elevation, vegetation cover and precipitation gradients: a case study in the Taihang Mountains. Ecol Indic 103:94–104. https://doi.org/10.1016/j.ecolind.2019.03.034

Liu S, Behm JE, Wan S, Yan J, Ye Q, Zhang W, Fu S (2021) Effects of canopy nitrogen addition on soil fauna and litter decomposition rate in a temperate forest and a subtropical forest. Geoderma. https://doi.org/10.1016/j.geoderma.2020.114703

Loba A, Sykula M, Kierczak J, Labaz B, Bogacz A, Waroszewski J (2020) In situ weathering of rocks or aeolian silt deposition: key parameters for verifying parent material and pedogenesis in the Opawskie Mountains—a case study from SW Poland. J Soils Sediments 20:435–451. https://doi.org/10.1007/s11368-019-02377-5

Lucas M, Schulüter S, Vogel H, Vetterlein D (2019) Soil structure formation along an agricultural chronosequence. Geoderma 350:61–72. https://doi.org/10.1016/j.geoderma.2019.04.041

Luo Y, Zhao X, Li Y, Liu X, Wang L, Wang X, Du Z (2021) Wind disturbance on litter production affects soil carbon accumulation in degraded sandy grasslands in semi-arid sandy grassland. Ecol Eng 171:106737. https://doi.org/10.1016/j.ecoleng.2021.106373

Ma X, Zarebanadkouki M, Kuzyakov Y, Blagodatskaya E, Pausch J, Razavi BS (2018) Spatial patterns of enzyme activities in the rhizosphere: effects of root hairs and root radius. Soil Biol Biochem 118:69–78. https://doi.org/10.1016/j.soilbio.2017.12.009

Maas GCB, Sanquetta CR, Marques R, Machado SA, Sanquetta MNI, Corte APD, Barberena IM (2021) Carbon production from seasonal litterfall in the Brazilian Atlantic Forest. Southern For J For Sci. https://doi.org/10.2989/20702620.2021.1886575

Machado DL, Pereira MG, Correia MEF, Diniz AR, Menezes CEB (2015) Fauna edáfica na dinâmica sucessional da mata atlântica em floresta estacional semidecidual na bacia do rio Paraíba do Sul – RJ. Cienc Florest 25:91–106. https://doi.org/10.5902/1980509817466

Magcale-Macandog DB, Manlubatan MBT, Edrial JJM, Mago KS, de Luna JEI, Nayoos J, Porcioncula RP (2018) Leaf litter decomposition and diversity of arthropod decomposers in tropical *Muyong* forest in Banaue, Philippines. Paddy Water Environ 16:265–277. https://doi.org/10.1007/s10333-017-0624-9

Mahmoud E, El-Beshbeshy T, El-Kader NA, El Shal R, Khalafallah N (2019) Impacts of biochar application on soil fertility, plant nutrients uptake and maize (*Zea mays* L.) yield in saline sodic soil. Arab J Geosci 12:719. https://doi.org/10.1007/s12517-019-4937-4

Malone B, Searle S (2020) Improvements to the Australian national soil thickness map using an integrated data mining approach. Geoderma 377:114579. https://doi.org/10.1016/j.geoderma.2020.114579

Manwaring M, Wallace HM, Weaver HJ (2018) Effects of a mulch layer on the assemblage and abundance of mesostigmatan mites and other arthropods in the soil of a sugarcane agro-ecosystem in Australia. Exp Appl Acarol 74:291–300. https://doi.org/10.1007/s10493-0180227-1

Marian F, Sandmann D, Krashevska V, Maraun M, Scheu S (2018) Altitude and decomposition stage rather than litter origin structure soil microarthropod communities in tropical montane rainforests. Soil Biol Biochem 125:263–274. https://doi.org/10.1016/j.soilbio.2018.07.017

Maru A, Haruna AO, Asap A, Majid NMA, Maikol N, Jeffary AV (2020) Reducing acidity of tropical acid soil to improve phosphorus availability and *Zea mays* L. productivity through efficient use of Chicken Litter Biochar and Triple Superphosphate. Appl Sci 10(6):2127. https://doi.org/10.3390/app10062127

Mauda EV, Joseph GS, Seymour CL, Munyai TC, Foord SH (2018) Changes in land use alter ant diversity, assemblage composition and dominant functional groups in African savannas. Biodivers Conserv 27:947–965. https://doi.org/10.1007/s10531-017-1474-x

McGee KM, Porter TM, Wright M, Hajibabaei M (2020) Drivers of tropical soil invertebrate community composition and richness across tropical secondary forests using DNA metasystematics. Sci Rep 10:e18429. https://doi.org/10.1038/s41598-020-75452-4

Meira-Neto JAA, Tolentino GS, da Silva CNA, Neri AV, Gastauer M, Magnago LFS, Yuste JC, Valladares F (2017) Functional antagonism between nitrogen-fixing leguminous trees and calcicole-drought-tolerant trees in the Cerrado. Acta Bot Brasil 31:11–18. https://doi.org/10.1590/0102-33062016abb0380

Melo LN, Souza TAF, Santos D (2019) Cover crop farming system affects macroarthropods community diversity in Regosol of Caatinga. Biologia, Brazil. https://doi.org/10.2478/s11756-019-00272-5

Mikutta R, Turner S, Schippers A, Gentsch N, Meyer-Stüve S, Condron LM, Pletze DA, Richardson SJ, Eger A, Hempel G, Kaiser K, Klotzbücher T, Guggenberger G (2019) Microbial and abiotic controls on mineral-associated organic matter in soil profiles along an ecosystem gradient. Sci Rep 9:10294. https://doi.org/10.1038/s41598-019-46501-4

Molina A, Vanacker V, Corre MD, Veldkamp E (2019) Patterns in soil chemical weathering related to topographic gradients and vegetation structure in a high Andean tropical ecosystem. J Geophys Res Earth 124:666–685. https://doi.org/10.1029/2018JF004856

Muchane MN, Sileshi GW, Gripenberg S, Jonsson M, Pumarinõ L, Barrios E (2020) Agroforestry boosts soil health in the humid and sub-humid tropics: a meta-analysis. Agric Ecosyst Environ 295:106899. https://doi.org/10.1016/j.agee.2020.106899

Myer A, Forschler BT (2019) Evidence for the role of subterranean termites (*Reticulitermes* spp.) in temperate forest soil nutrient cycling. Ecosystems 22:602–618. https://doi.org/10.1007/s10021-108-0291-8

Nakayama M, Imamura S, Taniguchi T, Tateno R (2019) Does conversion from natural forest to plantation affect fungal and bacterial biodiversity, community structure, and co-occurrence networks in the organic horizon and mineral soil? For Ecol Manag 446:238–250. https://doi.org/10.1016/j.foreco.2019.05.042

Nanganoa LT, Okolle JN, Missi V, Tueche JR, Levai LD, Njukeng JN (2019) Impact of different land-use systems on soil physicochemical properties and macrofauna abundance in the humid Tropics of Cameroon. Appl Environ Soil Sci. https://doi.org/10.1155/2019/5701278

Negasa T, Ketema H, Legesse A, Sisay M, Temesgen H (2017) Variation in soil properties under different land use types managed by smallholder farmers along the toposequence in southern Ethiopia. Geoderma 290:40–50. https://doi.org/10.1016/j.geoderma.2016.11.021

Ng K, McIntyre S, Macfadyen S, Barton PS, Driscoll DA, Lindenmayer DB (2018) Dynamic effects of ground-layer plant communities on beetles in a fragmented farming landscape. Biodivers Conserv 27:2131–2153. https://doi.org/10.1007/s10531-018-1526-x

Olatunji OA, Gong S, Tariq A, Pan K, Sun X, Chen W, Zhang L, Ak Dkil M, Huang D, Tan X (2019) The effect of phosphorus addition, soil moisture, and plant type on soil nematode abundance and community composition. J Soils Sediments 19:1139–1150. https://doi.org/10.1007/s11368-018-2146-5

Oliveira G, Carvalho MEA, Silva HF, Brignoni AS, Lima LR, Camargos LS, Souza LA (2021) *Lonchocarpus cultratus*, a Brazilian savanna tree, endures high soil Pb levels. Environ Sci Pollut Res 28:50931–50940. https://doi.org/10.1007/s11356-021-15856-5

Oyedeji S, Agboola OO, Animasaun DA, Ogunkunle CO, Fatoba PO (2021) Organic carbon, nitrogen and phosphorus enrichment potentials from litter fall in selected greenbelt species during a seasonal transition in Nigeria's savanna. Trop Ecol. https://doi.org/10.1007/s42965-021-00172-3

Pasley HR, Cairns JE, Camberato JJ, Vyn TJ (2019) Nitrogen fertilizer rate increases plant uptake and soil availability of essential nutrients in continuous maize production in Kenya and Zimbabwe. Nutr Cycl Agroecosyst 115:373–389. https://doi.org/10.1007/s10705-019-10016-1

Pastorelli R, Costagli V, Forte C, Viti C, Rompato B, Nannini G, Certini G (2021) Litter decomposition: little evidence of the "home-field advantage" in a mountain forest in Italy. Soil Biol Biochem 159:108300. https://doi.org/10.1016/j.soilbio.2021.108300

Paymaneh Z, Sarcheshmehpour M, Bukovská P, Jansa J (2019) Could indigenous arbuscular mycorrhizal communities be used to improve tolerance of pistachio to salinity and/or drought? Symbiosis 79:269–283. https://doi.org/10.1007/s13199-019-00645-z

Peng X, Wang X, Dai Q, Ding G, Li C (2020a) Soil structure and nutrient contents in underground fissures in a rock-mantled slope in the karst rocky desertification area. Environ Earth Sci 79:3. https://doi.org/10.1007/s12665-019-8708-z

Peng X, Wu W, Zheng Y, Sun J, Hu T, Wang P (2020b) Correlation analysis of land surface temperature and topographic elements in Hangzhou, China. Sci Rep 10:10451. https://doi.org/10.1038/s41598-020-67423-6

Pestana LFA, de Souza ALT, Tanaka MO, Lasbarque FM, Soares JAH (2020) Interactive effects between vegetation structure and soil fertility on tropical ground-dwelling arthropod assemblages. Appl Soil Ecol 155:103642. https://doi.org/10.1016/j.apsoil.2020.103624

Plaas E, Meyer-Wolfarth F, Banse M, Bengtsson J, Bergmann H, Faber J, Potthoff M, Runge T, Schrader S, Taylor A (2019) Towards valuation of biodiversity in agricultural soils: a case for earthworms. Ecol Econ 159:291–300. https://doi.org/10.1016/j.ecolecon.2019.02.003

Potatov AM, Goncharov AA, Semenina EE, Korotkevich AY, Tsurikov SM, Rozanova OL, Anichkin AE, Zuey AG, Samoylova ES, Semenyuk II, Yevdokimov IV, Tiunov AV (2017) Arthropods in the subsoil: abundance and vertical distribution as related to soil organic matter, microbial biomass and plant roots. Eur J Soil Biol 82:88–97. https://doi.org/10.1016/j.ejsobi.2017.09.001

Purwanto BH, Alam S (2020) Impact of intensive agricultural management on carbon and nitrogen dynamics in the humid tropics. Soil Sci Plant Nutr 66:50–59. https://doi.org/10.1080/0038076 8.2019.1705182

Queiroz MG, da Silva TGF, Zolnier S, de Souza CAA, de Souza LSB, Steidle Neto AJ, de Araújo GGL, Ferreira WPM (2019) Seasonal patterns of deposition litterfall in a seasonal dry tropical forest. Agric Forest Meteorol 279:107712. https://doi.org/10.1016/j.agrformet.2019.107712

Rahbek C, Borregaard MK, Antonelli A, Colwell RK, Holt BG, Nogues-Bravo D, Rasmussen CMØ, Richardson K, Rosing MT, Whittaker RJ, Fjeldså J (2019) Building mountain biodiversity: geological and evolutionary processes. Science 365:1114–1119. https://doi.org/10.1126/science.aax0151

Raiesi F, Salek-Gilani S (2020) Development of a soil quality index for characterizing effects of land-use changes on degradation and ecological restoration of rangeland soils in a semi-arid ecosystem. Land Degrad Dev. https://doi.org/10.1002/ldr.3553

Raj JK (2018) Physical characterization of a deep weathering profile over rhyolite in humid tropical peninsular Malaysia. Geotech Geol Eng 36:3793–3809. https://doi.org/10.1007/s10706-018-0572-1

Rajapaksha RPSK, Karunatne SB, Biswas A, Paul A, Madawala HMSP, Gunathilake SK, Ratnayake RR (2020) Identifying the spatial drivers and scale-specific variations of soil organic carbon in tropical ecosystems: a case study from Knuckles Forest Reserve in Sri Lanka. For Ecol Manag 474:118285. https://doi.org/10.1016/j.foreco.2020.118285

Ramírez BH, van der Ploeg M, Teuling AJ, Ganzeveld L, Leemans R (2017) Tropical Montane Cloud Forests in the Orinoco river basin: the role of soil organic layers in water storage and release. Geoderma 298:14–26. https://doi.org/10.1016/j.geoderma.2017.03.007

Rasmussen PU, Bennett AE, Tack AJM (2019) The impact of elevated temperature and drought on the ecology and evolution of plant-soil microbe interactions. J Ecol. https://doi.org/10.1111/1365-2745.13292

Real-Santillán RO, del Val E, Cruz-Ortega R, Contreras-Cornejo HÁ, González-Esquivel CE, Larsen J (2019) Increased maize growth and P uptake promoted by arbuscular mycorrhizal fungi coincide with higher foliar herbivory and larval biomass of the fall armyworm *Spodoptera frugiperda*. Mycorrhiza 29:615–622. https://doi.org/10.1007/s00572-019-00920-3

Rodríguez J, Thompson V, Rubido-Bará M, Cordero-Rivera A, González L (2019) Herbivore accumulation on Invasive alien plants increases the distribution range of generalist herbivorous insects and supports proliferation of non-native insect pests. Biol Invasions 21:1511–1527. https://doi.org/10.1007/s10530-019-01913-1

Rodriguez-Caballero E, Belnap J, Büdel B, Crutzen PJ, Andreae MO, Pöschl U, Weber B (2018) Dryland photoautotrophic soil surface communities endangered by global change. Nat Geosci 11:185–189. https://doi.org/10.1038/s41561-018-0072-1

Roy S, Roy MM, Jaiswal AK, Baitha A (2018) Soil arthropods in maintaining soil health: thrust areas for sugarcane production systems. Sugar Tech 20:376–391. https://doi.org/10.1007/s12355-018-0591-5

Sadiq FK, Maniyunda LM, Anumah AO, Adegoke KA (2021) Variation of soil properties under different landscape positions and land use in Hunkuyi, Northern Guinea savanna of Nigeria. Environ Monit Assess 193:178. https://doi.org/10.1007/s10661-021-08974-7

Sahu PK, Singh DP, Prabha R, Meena KK, Abhislash PC (2019) Connecting microbial capabilities with the soil and plant health: options for agricultural sustainability. Ecol Indic 105:601–612. https://doi.org/10.1016/j.ecolind.2018.05.084

Saifullah DS, Naeem A, Rengel Z, Naidu R (2018) Biochar application for the remediation of salt-affected soils: challenges and opportunities. Sci Total Environ 625:320–335. https://doi.org/10.1016/j.scitotenv.2017.12.257

Santos AC, da Silva RC, da Silva Neto EC, dos Anjos LHC, Pereira MG (2021) Weathering and pedogenesis of mafic rock in the Brazilian Atlantic Forest. J S Am Earth Sci 111:103452. https://doi.org/10.1016/j.jsames.2021.103452

Santos-Heredia C, Andresen E, Zárate DA, Escobar F (2018) Dung beetles and their Ecological functions in three agroforestry systems in the Lacandona rainforest of Mexico. Biodivers Conserv 27:2379–2394. https://doi.org/10.1007/s10531-018-1542-x

Sato T, Hachiya S, Inamura N, Ezawa T, Cheng W, Tawaraya K (2019) Secretion of acid phosphatase from extraradical hyphae of the arbuscular mycorrhizal fungus *Rhizophagus clarus* is regulated in response to phosphate availability. Mycorrhiza 29:599–605. https://doi.org/10.1007/s00572-019-00923-0

Silva RA, Siqueira GM (2020) Multifractal analysis of soil fauna diversity. Bragantia Indexes. https://doi.org/10.1590/1678-4499.20190179

Singh S, Mayes MA, Shekoofa A, Kivlin SN, Bansal S, Jagadamma S (2021) Soil organic carbon cycling in response to simulated soil moisture variation under field conditions. Sci Rep 11:10841. https://doi.org/10.1038/s41598-021-90359-4

Smagin AV, Sadovnikova NB, Kirichenko AV, Egorov YV, Vityazev VG, Bashina AS (2018) Dependence of the osmotic pressure and electrical conductivity of soil solutions on the soil water content. Eurasian Soil Sci 51:1462–1473. https://doi.org/10.1134/S1064229318120128

Sofo A, Mininni AN, Ricciuti P (2020) Comparing the effects of soil fauna on litter decomposition and organic matter turnover in sustainably and conventionally managed olive orchards. Geoderma 372:114393. https://doi.org/10.1016/j.geoderma.2020.114393

Sousa NMF, Veresoglou SD, Oehl F, Rillig MC, Maia LC (2018) Predictors of arbuscular mycorrhizal fungal communities in the Brazilian Tropical Dry Forest. Soil Microbiol 75:447–458. https://doi.org/10.1007/s00248-017-1042-7

Souza TAF, Freitas H (2018) Long-Term effects of fertilization on soil organism diversity. In: Gaba S, Smith B, Lichtfouse E (eds) Sustainable agriculture reviews 28. Sustainable agriculture reviews. Springer, Cham. https://doi.org/10.1007/978-3-319-90309-5_7

Souza TAF, Rodrígues AF, Marques LF (2016) Long-term effects of alternative and conventional fertilization on macroarthropod community composition: a field study with wheat (*Triticum aestivum* L) cultivated on a Ferralsol. Org Agric 6:323–330. https://doi.org/10.1007/s13165-015-0138-y

Souza TAF, Santos (2018) Biologia dos Solos da Caatinga. Universidade Federal da Paraíba, PPGCS, Areia

Stoops G, Langohr R, Van Ranst E (2020) Micromorphology of soils and palaeosoils in Belgium. An inventory and meta-analysis. Catena 194:104718. https://doi.org/10.1016/j.catena.2020.104718

Suleiman MK, Dixon K, Commander L, Nevill P, Quoreshi AM, Bhat NR, Manuvel A, Sivadasan MT (2019) Assessment of the Diversity of fungal community composition associated with *Vachellia pachyceras* and its rhizosphere soil from Kuwait desert. Front Microbiol. https://doi.org/10.3389/fmicb.2019.00063

Sun T, Wang Y, Hui D, Jing X, Feng W (2020a) Soil properties rather than climate and ecosystem type control the vertical variations of soil organic carbon, microbial carbon, and microbial quotient. Soil Biol Biochem 148:107905. https://doi.org/10.1016/j.soilbio.2020.107905

Sun Y, Luo C, Jiang L, Song M, Zhang D, Li J, Li Y, Ostle NJ, Zhang G (2020b) Land-use changes alter soil bacterial composition and diversity in tropical forest soil in China. Sci Total Environ. https://doi.org/10.1016/j.scitotenv.2020.136526

Swift MJ, Heal OW, Anderson JM (1979) Decomposition in terrestrial ecosystems. University of California Press, Berkeley

Tabrizi RA, Dontsova K, Grachet NG, Tfaily MM (2022) Elevated temperatures drive abiotic and biotic degradation of organic matter in a peat bog under oxic conditions. Sci Total Environ 804:150045. https://doi.org/10.1016/j.scitotenv.2021.150045

Thakur MP, van der Putten WH, Wilschut RA, Veen GFC, Kardol P, van Ruiiven J, Allan E, Roscher C, van Kleunen M, Bezemer TM (2021) Plant–soil feedbacks and temporal dynamics of plant diversity–productivity relationships. Trends Ecol Evol 36:651–661. https://doi.org/10.1016/j.tree.2021.03.011

Tunlid A, White DC (1992) Biochemical analysis of biomass, community structure, nutritional status, and metabolic activity of microbial communities in soil. In: Stotzky G, Bollag JM (eds) Soil biochemistry, vol 7, Marcel Dekker, NY. https://doi.org/10.1201/9781003210207-7

Velasquez E, Lavelle P (2019) Soil macrofauna as an indicator for evaluating soil based ecosystem services in agricultural landscapes. Acta Oecol 100. https://doi.org/10.1016/j.actao.2019.103446

Wale M, Yesuf S (2021) Abundance and diversity of soil arthropods in disturbed and undisturbed ecosystem in Western Amhara, Ethiopia. Int J Tropi Insect Sci. https://doi.org/10.1007/s42690-021-00600-w

Wei W, Zhang S, Wu L, Cui D, Ding X (2020) Biochar and phosphorus fertilization improved soil quality and inorganic phosphorus fractions in saline-alkaline soils. Arch Agron Soil Sci 67:177–1190. https://doi.org/10.1080/03650340.2020.1784879

Wietrzyk-Pelka P, Rola K, Szymański W, Wegrzyn MH (2020) Organic carbon accumulation in the glacier forelands with regard to variability of environmental conditions in different ecogenesis stages of High Arctic ecosystems. Sci Total Environ 717:135151. https://doi.org/10.1016/j.scitotenv.2019.135151

Woese CR (1987) Bacterial evolution. Microbiol Rev 51:221–271

Wood TE, Cavaleri MA, Giardina CP, Khan S, Mohan JE, Nottingham AT, Reed SC, Slot M (2019) Soil warming effects on tropical forests with highly weathered soils. Ecosyst Conseq Soil Warm 385–439. https://doi.org/10.1016/B978-0-12-813493-1.00015-6

Xu M, Kang S, Wu H, Yuan X (2018) Detection of spatio-temporal variability of air temperature and precipitation based on long-term meteorological station observations over Tianshan Mountains, Central Asia. Atm Res 203:141–163. https://doi.org/10.1016/j.atmosres.2017.12.007

Xu S, Sayer EJ, Eisenhaeuer N, Lu X, Wang J, Liu C (2021) Aboveground litter inputs determine carbon storage across soil profiles: a meta-analysis. Plant Soil 462:429–444. https://doi.org/10.1007/s11104-021-04881-5

Yang B, Zhang W, Xu H, Wang S, Xu X, Fan H, Cheng HYH, Ruan H (2018) Effects of soil fauna on leaf litter decomposition under different land uses in eastern coast of China. J For Res 29(4):973–982. https://doi.org/10.1007/s11676-017-0521-5

Yu F, Hunt AG (2018) Predicting soil formation on the basis of transport-limited chemical weathering. Geomorphology 301:21–27. https://doi.org/10.1016/j.geomorph.2017.10.027

Zhang K, Maltais-Landry G, Liao H (2021) How soil biota regulate C cycling and soil C pools in diversified crop rotations. Soil Biol Biochem 156:108219. https://doi.org/10.1016/j.soilbio.2021.108219

Zhang W, Yang K, Lyu Z, Zhu J (2019) Microbial groups and their functions control the decomposition of coniferous litter: A comparison with broadleaved tree litters. Soil Biol Biochem 133:196–207. https://doi.org/10.1016/j.soilbio.2019.03.009

Zhang X, Kuzyakov Y, Zang H, Dippold MA, Shi L, Spielvogel S, Razavi BS (2020) Rhizosphere hotspots: Root hairs and warming control microbial efficiency, carbon utilization and energy production. Soil Biol Biochem 148:107872. https://doi.org/10.1016/j.soilbio.2020.107872

Zhao Y, Wang S, Li Y, Liu J, Zhuo Y, Chen H, Wang J, Xu L, Sun Z (2018) Extensive reclamation of saline-sodic soils with flue gas desulfurization gypsum on the Songnen Plain, Northeast China. Geoderma 321:52–60. https://doi.org/10.1016/j.geoderma.2018.01.033

Zheng W, Zhao Z, Gong Q, Zhai B, Li Z (2018) Responses of fungal–bacterial community and network to organic inputs vary among different spatial habitats in soil. Soil Biol Biochem 125:54–63. https://doi.org/10.1016/j.soilbio.2018.06.029

Zhou Z, Chuankuan Z, Luo WY (2018) Response of soil microbial communities to altered precipitation: A global synthesis. Glob Ecol Biogeogr 27(9):1121–1136. https://doi.org/10.1111/geb.12761

Zou Z, Xiao X, Zhou H, Chen Z, Zeng J, Wang W, Feng G, Huang X (2018) Effects of soil acidification on the toxicity of organophosphorus pesticide on *Eisenia fetida* and its mechanism. J Hazard Mater 359:365–372. https://doi.org/10.1016/j.jhazmat.2018.04.036

Zuo Y, He C, He X, Li X, Xue Z, Li X, Wang S (2019) Plant cover of Ammopiptanthus mongolicus and soil factors shape soil microbial community and catabolic functional diversity in the arid desert in Northwest China. Appl Soil Ecol. https://doi.org/10.1016/j.apsoil.2019.103389

# Chapter 2
# The Living Soil

**Abstract** In the previous chapter, the tropical soil types and their forming factors and inner physicochemical properties were introduced. The next issue to be considered is the living soil that is constituted by a wide diversity of soil organisms. Also, the soil biota classification based on their body size and taxonomic and functional groups is introduced within this chapter. The soil ecologists must consider the structure of the soil food web and the ecosystem services provided by soil organisms into tropical ecosystems. Basically, soil organisms are both above- and belowground individuals which might be living in soil ecosystem. They are difficult to understand and to study because of their vast diversity of taxonomic groups. In tropical ecosystem, soil organisms build a complex soil food web which provides ecosystem services (e.g., soil organic matter formation, nutrient cycling, bioturbation, and control of pests and diseases). Finally, this chapter will introduce the relationships among soil organisms and how do they are influenced by environmental disturbances.

**Keywords** Diversity of soil organisms · Macrofauna · Mesofauna · Microbiota · Soil food web

**Questions Covered in the Chapter**
1. How can we classify soil organisms?
2. Describe the main biological properties of the soil ecosystem.
3. What does the expression "soil organism" means?
4. How many functional groups can we find?
5. How many services do the soil organisms provide into the soil profile? Describe all of them.
6. How can we define the soil food web?
7. How are soil organisms influenced both at temporal and spatial scale?
8. Describe the methods to identify soil organisms? Why are they important for soil ecology or soil biology?
9. Why is it important to study soil organism groups?

© The Author(s), under exclusive license to Springer Nature Switzerland AG 2022          23
T. Souza, *Soil Biology in Tropical Ecosystems*, https://doi.org/10.1007/978-3-031-00949-5_2

## 2.1  Introduction

In tropical ecosystems, many researchers have reported a vast diversity of organisms (e.g., arachnids, bacteria, fungi, insects, myriapods, nematodes, protozoans, snails, and springtails) living above- and belowground soil profile (Gebremikael et al. 2016; Bai et al. 2018; Melo et al. 2019; Anyango et al. 2020; Forstall-Sosa et al. 2020). We can classify these soil individuals according to their body size (e.g., generalist classifications), morphological structures (e.g., specific taxonomical level), and function or even services (e.g., quite common in soil ecology studies) (Swift et al. 1979; Souza and Freitas 2018 Souza and Santos 2018). According to Silva and Siqueira (2020), the living soil presents an extremely complex food web which is hard to understand and to study. Also, the little effort and lack of studies of belowground organisms when compared with aboveground organisms could be considered surprising if we consider that most soil organism species might be living in soil (Fig. 2.1).

At the tropics, we have old acid soils (e.g., Oxisols) characterized by high contents of $Al^{3+}$ and low fertility (Machado et al. 2015; Souza et al. 2016; Manwaring et al. 2018), and the soil organisms into this context regulate important ecosystem processes, such as soil nutrient availability, biological control, soil organic matter transformation, soil structure, and primary production (Basirat et al. 2019; Liu et al. 2019a, b; Paymaneh et al. 2019). They also provide biological properties to soil profile such as metabolism (e.g., if we consider the soil enzymatic activity, rhizodeposition, and root exudation), soil respiration (e.g., by biota and microbiota activity), evolution (e.g., if we consider soil organisms as a soil formation a factor), and energy supply (e.g., thought the breakdown of litter, layers L and H, and soil organic matter formation) for soil organisms, plants, and human being (Gebremikael et al. 2016; Real-Santillán et al. 2019; Sato et al. 2019; Anyango et al. 2020).

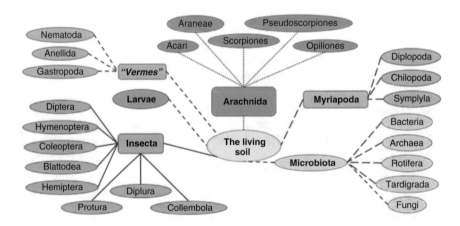

**Fig. 2.1** Schematic view of the aboveground and belowground diversity into tropical soil ecosystems. (Adapted from Souza and Santos (2018))

## 2.2 The Soil Organisms

Even the expression "the soil organisms" be very direct and objective, when it describes each discrete living thing (e.g., plants, insects, and microorganisms) which presents biological properties and activities (e.g., evolution, respiration, reproduction, and metabolism) into soil ecosystem (Balota 2017). It provides a high variability of taxonomic groups whose primary role is to promote ecosystem services and soil quality (Ng et al. 2018), and we can define this variability of "living things" as the soil food web (Bardgett 2005). We can classify this vast variability of living things according to their body size, taxonomical characteristics, and functional groups (Fig. 2.2).

According to Swift et al. (1979), we can split the entire soil food web in three main groups (e.g., macrofauna, mesofauna, and microfauna) based in the body width of each single organism. The macrofauna group includes soil organisms with body width larger than 2.0 mm. Next, the mesofauna group includes soil organisms with body width varying between 0.1 mm and 2.0 mm, and finally the microfauna group includes all soil organisms with body width below 0.1 mm. This last one comprises the most numerous and diverse individuals (e.g., Acari, Bacteria, Fungi, Nematoda, Protozoa, and Rotifera), accordingly Swift's body size classification (Fig. 2.2a). This classification is generalist, and it only describes major groups without any ecological relationship or even function. It helps the beginners during their field studies, and it is particularly useful when we are aiming to define specific groups to start a field collection (e.g., Microfauna – Fungi – Glomeromycota – *Funneliformis* or Macrofauna – Insecta – Coleoptera – Scarabaeidae).

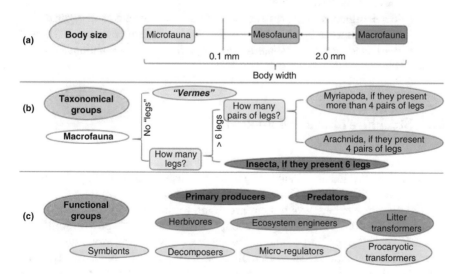

**Fig. 2.2** Soil organism classification following their body size (**a**) and taxonomical (**b**) and functional groups (**c**). (Adapted from Swift et al. (1979), Souza and Freitas (2018), and Souza and Santos (2018))

On the other hand, we can classify each soil organism based on its shared characteristics (e.g., we can classify soil organisms into the Macrofauna major group only counting the number of legs in each collected soil organism following Souza and Santos' classification) (Fig. 2.2b). We can group soil organisms together into taxa. It is hard to perform at specific taxonomic ranks (e.g., genus and even species), but easy to perform at high rank levels (e.g., Order and Family levels). In both situations, students need practice, and they may use key to classify organisms and to perform courses to improve their skills during identification of each aimed group (e.g., springtails, spiders, bugs, etc.). However, this second approach to identify soil organisms only provides a taxa list without any function, ecological process, or trophic level. Nowadays, with the advent of such fields such as molecular biology, we can use molecular tools (e.g., phospholipid fatty acid analysis, polymerase chain reaction, and gel electrophoresis) to identify the true range of soil organisms at species level (Woese 1987; Tunlid and White 1992; Kowalchuck et al. 1997).

So, we need more if we are trying to understand the role of soil organisms in recycling of organic matter from the aboveground in tropical ecosystem. Functional group classification (Souza and Freitas 2018) provides a clear and concise identification grouping soil organisms according to their function above- and belowground ecosystem (Fig. 2.2c). These authors define nine functional groups (e.g., decomposers, ecosystem engineers, herbivores, litter transformers, micro-regulators, predators, primary producers, prokaryotic transformers, and symbionts) based on their morphological, physiological, behavioral, and biochemical characteristics and their services to the ecosystem. Using functional groups classification, we can respond the following questions:

1. How does soil ecosystem work?
2. How do soil organisms are providing ecosystem services (e.g., soil organic matter formation, nutrient cycling, bioturbation, and control of pests and diseases)?
3. How does land use or seasonality change soil organism functional group's abundance?

## 2.3  Structure of the Soil Food Web

The soil organisms comprise a wide range of trophic levels (e.g., varying in feed specialization, life history strategies, and spatial distribution) creating a complex soil food web (Fig. 2.3). Into this web, they are allocated based on their functional group. Other researchers group them according to their feeding habit and living habitat (Swift et al. 1979; Bardgett 2005; Gebremikael et al. 2016). Some soil organisms act aboveground promoting biological control, herbivory, litter deposition, and transformation (Santos-Heredia et al. 2018; Yang et al. 2018; Myer and Forschler 2019; Rodriguez et al. 2019), whereas others act belowground promoting biological control, decomposition, mineralization, nutrient cycling, passive transport of symbionts (e.g., earthworms help arbuscular mycorrhizal fungi spread into

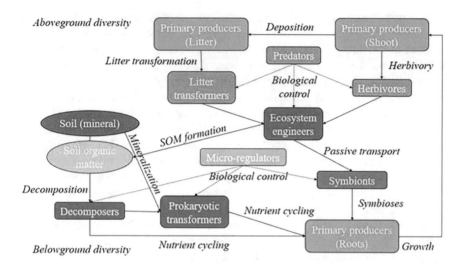

**Fig. 2.3** Summarized structure of the soil food web. (Adapted from Ng et al. (2018), Souza and Freitas (2018), Souza and Santos (2018), Yang et al. (2018))

new habitats), plant growth, soil organic matter formation, and symbioses with host plants (Lammel et al. 2015; Carrilo-Saucedo et al. 2018; de Novais et al. 2020). A complex and robust soil food web (e.g., with high richness, diversity, and abundance of functional groups) may indicate high-quality sites (e.g., by high provision of habitat) and environmental sustainability (e.g., by high provision of energy supply) as described by Melo et al. (2019).

The primary producers start growing above- and belowground, and in old tissues, the senescence process is the main pathway which improves litter deposition. It is modulated by plant genotype, plant community composition (e.g., litter quantity and quality), physiological pathways (e.g., plant nutrient and hormone balance), climate changes (e.g., some plants in tropical ecosystem lose their leaves during dry periods as a strategy against drought), and herbivory activity (e.g., by improving phytohormone release and plant growth) (Real-Santillán et al. 2019). The whole litter[1] deposited aboveground has two main primary functions on soil food web: (i) it serves as habitat to a high variety of soil organisms, such as litter transformers, herbivores, and ecosystem engineers (Melo et al. 2019), and (ii) it acts as energy supply (food) to litter transformers (Ng et al. 2018). On the other hand, herbivores feed fresh plant tissue (e.g., leaves). In both groups, litter transformers and herbivores are responsible for breaking down (e.g., by their physical trituration) litter and fresh plant tissues through their mouthparts which present a robust pair of mandibles (Coyle et al. 2017; Santos-Heredia et al. 2018). These two functional groups create favorable conditions (e.g., by providing habitat and food) for ecosystem engineers. Ecosystem engineers (e.g.,

---

[1] According to Souza and Santos (2018), the whole litter is characterized by plant and animals' residues deposited on soil surface with mean diameter less than 2 mm.

ants, isopods, termites, and wild cockroach) are the most abundant individuals of macrofauna group which creates nests and galleries into soil profile promoting soil structure and creating a whole habitat to sustain a complex belowground soil food web (Myer and Forschler (2019). This group also plays three important processes into soil profile: (i) soil organic matter formation by helping litter transformers mixing organic residues, (ii) promoting passive transport of arbuscular mycorrhizal spores and N-fixing bacteria cells, and (iii) promoting seed bank exhumation (Santos-Heredia et al. 2018). The positive effects promoted by litter transformers, herbivores, and ecosystem engineers on plant-soil interface are well-known (Jo et al. 2020), but the abundance and occurrence of individuals of these three groups must be controlled by predators to avoid population problems (Balkenhol et al. 2018). The complex relationship among these five groups (e.g., ecosystem engineers, herbivores, litter transformers, predators, and primary producers) comprises the aboveground diversity (Souza and Freitas 2018). They also are included inside the macrofauna group (According Swift's classification), and the monitoring of their abundance, richness, diversity, and dominance could be used as bioindicators[2] for changes in land use (e.g., C and N pools), soil and water quality (e.g., presence of toxic elements), and environmental hardening (Kamau et al. 2017; Mauda et al. 2018).

Before presenting the ecosystem services belowground promoted by soil food web, we must consider the soil ecosystem as two dissimilar and well-connected phases:

1. Mineral phase: It comprises 45% of soil ecosystem, and it is formed by the combined effect of physical, chemical, and biological processes working on soil parent material.
2. Organic phase: It comprises 5% of soil ecosystem, and it is formed by litter deposition and transformation promoted by the soil macrofauna group (e.g., ecosystem engineers, litter transformers, and herbivores) and added into soil by organic matter formation process.

Considering each soil phase separately, we should emphasize that only prokaryotic transformers (e.g., archaea, bacteria, and fungi) explore the microhabitat of the soil mineral phase, promoting the mineralization and plant nutrient release (e.g., contributing to nutrient cycling process) (Liu et al. 2019a, b). It may be considered as the shortest pathway where plant acquires nutrient to improve their own growth (Raiesi and Salek-Gilani 2020), and this pathway is modulated by soil parental material (e.g., residuum, aeolian, alluvium, lacustrine, marine deposits, glacial deposits, and colluvium), soil type (e.g., Arenosols, Cambisols, Ferralsols, and Regosols), and soil management (Rasmussen et al. 2019).

The main function of the soil organic phase (e.g., soil organic matter) is to provide resources to decomposers which breakdown complex organic substances and xenobiotic compounds and provide available nutrients for plants and prokaryotic transformers through a vast array of soil enzymes (Liu et al. 2019b). It improves soil

---

[2] We can extract and identify these soil organisms by simple approaches. See Sect. 2.5 to more details about how to extract and identify soil macrofauna using Provid-type traps.

nutrient cycling process and promotes primary producers by increasing their root biomass production and in turn plant growth (Doniger et al. 2020). In well-conserved tropical ecosystems, the entire soil food web acts promoting soil fertility, soil quality, net primary production, and water quality (Melo et al. 2019; Forstall-Sosa et al. 2020; Heydari et al. 2020). Despite its importance to the functioning of tropical ecosystems, we must consider that changes in soil food web structure, habitat, and energy supply to soil organisms can affect ecological processes and create negative plant-soil feedback (e.g., when decomposers, such as Bacteria and Fungi, act as plant pathogens). According to Souza and Freitas (2018), there are four pathways which described how do soil organisms respond to soil disturbances:

1. Unaltered community: If any natural or anthropic disturbance occurs spatially or temporally and the soil contents of C and N remain the same, we must find an unaltered soil biota community. It is common in tropical moist forest ecosystems.
2. Resilient community: Under seasonal disturbances (e.g., drought and food supply to soil organisms) which may change temporally both contents of C and N, the soil biota abundance could change temporally too, but after few days/months, it returns to its original structure (e.g., diversity, dominance, richness, and abundance). Here, the soil biota community (e.g., in general microfauna groups – Protozoa and Nematodes) is resilience to the same disturbances, and we can find this phenomenon in arid and semiarid ecosystems.
3. Resistant community: Under spatial/local disturbances (e.g., fire, hurricanes, cyclones, hail, and frost) which may alter significantly aboveground diversity, but with no changes on soil C and N contents, the soil biota community is still unaltered. Here, the soil biota community is resistance to natural disturbances, and we can find this phenomenon in tropical island ecosystems.
4. Alien community: Under long-term disturbances, the levels of C and N decrease significantly into soil profile. It changes permanently the original soil community structure, opening a functional redundancy process. Here, we can find two basic paths: (i) an alien community performs the same ecosystem services as the original did (e.g., Myriapods promoting litter transforming instead beetles) and (ii) the alien community performs differently as the original did (e.g., *Pseudomonas* starts causing plant diseases instead of promoting soil organic matter decomposition).

## 2.4  Patterns of Soil Biodiversity

In the tropics, the biological diversity varies in patterns across and into ecosystems (e.g., tropical rainforests, seasonal tropical forest, dry forests, spiny forests, desert, and islands) (Sousa et al. 2018; Suleiman et al. 2019; Doniger et al. 2020; Forstall-Sosa et al. 2020; Heydari et al. 2020; Sun et al. 2020). We can find an infinite variation in patterns of soil biota composition and structure which varies constantly according to seasonality and at spatial (e.g., plant community composition, and plant diversity) and temporal scales (e.g., plant community structure) (Fig. 2.4).

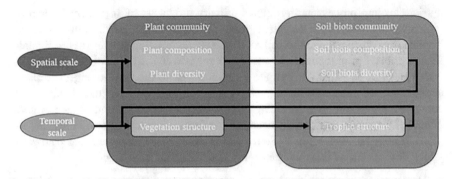

**Fig. 2.4** Patterns of soil biodiversity at temporal and spatial scale. (Adapted from Ng et al. (2018))

At spatial scale, tropical ecosystems may affect the patterns of soil biota diversity by changing soil organism composition and diversity as described by the habitat enemies and resource concentration hypotheses (Melo et al. 2019). In this case, there are stronger effects of plant species composition and plant diversity on herbivores and litter transformers (Heydari et al. 2020). On the other hand, at temporal scale, the relationship between plant-soil biota varies over time (Sun et al. 2020), and vegetation structure might change trophic structure of soil biota by affecting predator's abundance (Bai et al. 2018; Balkenhol et al. 2018). Thus, the combined effects of both spatial and temporal scale will vary according to habitat type, showing stronger effects of plant species composition on soil biota composition in some tropical ecosystem types (Sun et al. 2020).

Patterns of spatial variation in soil biota diversity have been explored, but perhaps the most prominent and intensive reports are into tropical ecosystems and its related soil biota diversity and richness (Table 2.1). Even if tropics are being recognized as global high-diversity center (Bardgett 2005), it has long been recognized that both diversity and richness of soil biota are lowest at low plant diversity ecoregion (e.g., desert, and dry forest). Plant community provides habitat and energy to soil biota (Ng et al. 2018; Silva and Siqueira 2020).

Other factors (e.g., topography, soil fertility, soil moisture, soil organic carbon and nitrogen, drought, biological invasion, and vegetation patchiness) could affect the patterns of soil biota at the landscape level (Lazzaro et al. 2018; Tajik et al. 2020; Lucena et al. 2021; Nsengimana et al. 2021). Even with an almost infinite variation in patterns of soil biota diversity across tropical ecosystems, we can assume some important points before starting our studies in soil ecology or soil biology:

1. Ecosystem net primary production is a determinant to soil biodiversity.
2. Patterns of soil biota diversity change constantly over long timescales because of succession process.
3. At short timescales such as seasonality changes (e.g., air temperature, soil moisture, etc.) promoted by the seasons of the year, just the soil biota abundance changes. There is no significant on soil biota diversity.

**Table 2.1** Patterns of soil biota diversity on tropical ecosystems

| Reference | Tropical ecosystem | Soil biota group | Diversity (Shannon's index) | Richness |
|---|---|---|---|---|
| Suleiman et al. (2019), Doniger et al. (2020) | Desert | Microbiota | 1.7[a] | 9 |
| Sousa et al. (2018) | Dry forest | Microbiota (Mycorrhizas) | 3.0 | 11 |
| Heydari et al. (2020) | Seasonal dry forest | Mesofauna | 1.5 | 6 |
| Lira et al. (2020) | Seasonal dry forest | Macrofauna | 2.1 | 11 |
| Dial et al. (2006) | Spiny forests | Macrofauna | 2.0 | 8 |
| Sun et al. (2020) | Islands | Microbiota (Bacteria) | 6.3 | – |
| Forstall-Sosa et al. (2020) | Tropical rainforests | Macrofauna | 3.5 | 15 |

[a]Data on each cell represent the mean values observed in each used reference

4. At the tropics, the number of macrofauna taxonomical units is highest and decreases towards the poles.
5. Soil organisms of the microfauna group are cosmopolitan such as arbuscular mycorrhizal fungi and other microbes, thus being able to migrate across the Earth's surface.
6. Mull humus soils tend to support a higher diversity of soil organisms than humus soils.
7. Soil biodiversity below plant species and around rhizosphere is higher than in adjacent exposed soils.
8. Soil biota community structure may be modulated by soil organic carbon, salts, and soil moisture content at spatial scale.
9. Ecosystem engineers (e.g., earthworms) are extremely sensitive to soil physical disturbances.
10. Organic horizons present a more active and diverse soil biota community than mineral horizons.
11. Mixtures of litters support a more diverse microfauna community than single-species litters.

## 2.5  Identifying Soil Biota Functional Groups

To identify the entire soil food web considering all soil organism groups at a specific taxonomic level, we must use a molecular approach combined with taxonomical parameters. It could be the most ideal scenario for soil ecologists. Molecular protocols to identify soil organisms are extremely specific and *primer* dependent. Also, these approaches give results at high cost and need a well-trained professional to read the outputs using software for molecular studies (e.g., BLAST, ContigExpress, etc.).

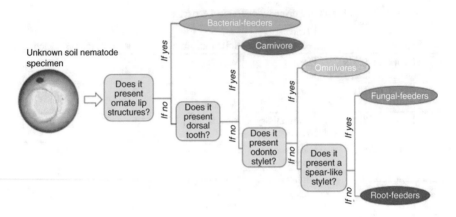

**Fig. 2.5** Key to classify soil nematodes based on their food behavior. (Adapted from Sechi et al. (2018), Mirsam et al. (2020))

On the other hand, to do a taxonomical classification, the young researcher needs practice and knowledge, and nowadays taxonomists are becoming a rare species into our scientific labs around the world. Even having the better conditions to run a molecular protocol and a taxonomical classification, we need to know what soil organism group we are aiming to study. Without a previous knowledge, the procedure will become stressful and will lead to a waste of money and time by the research group.

Thus, the firsts step is to identify the desired soil organism group and how the specimens from these groups are characterized. For example, soil nematodes are group together within five main groups based on their food behavior (Fig. 2.5). They can be classified as bacterial-feeder, carnivores, fungal-feeder, omnivores, and root-feeder nematodes just based on their lip morphology and the stylet type (e.g., odonto, spear-like, or curved stylets). According to Souza and Santos (2018), about 20,000 nematode species have been described so far, but the estimate of how many species exist in total indicates a number near to one million species. Once identifying the nematode group of interest, the researcher must select the correct kit to extract DNA and the specific primer for an aimed group, such as root-feeder nematodes (e.g., *Rhabditis*, *Meloidogyne*, *Pratylenchus*, and *Tylenchus* genus).

Other simple way to identify soil organism is when we use their morphotypes, such as the spore morphology of arbuscular mycorrhizal fungi (AMF). Spores from Glomeromycota can be classified as gigasporoid (e.g., *Cetraspora*, *Gigaspora*, *Dentiscutata*, *Quatunica*, *Racocetra*, and *Scutellospora* genus), entrophosporoid (e.g., *Entrophospora* genus), acaulosporoid (e.g., *Acaulospora*, *Ambispora*, *Archaeospora*, and *Otospora* genus), and glomoid (e.g., *Claroideoglomus*, *Funneliformis*, *Glomus*, *Paraglomus*, and *Rhizoglomus*) (Fig. 2.6). The AMF spore morphotype knowledge is the first step into the AMF identification. Also, this simple task can help when choosing specific primers for molecular classification as already described before for soil nematodes.

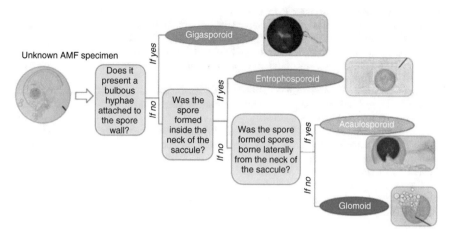

**Fig. 2.6** Key to classify arbuscular mycorrhizal spores based on their morphotypes. (Adapted from Souza and Freitas (2018))

## 2.6 Conclusions

In tropical ecosystem, there is a high diversity of soil biota that lives above- and belowground soil, and this diversity is modulated by plant community diversity, soil moisture, and favorable conditions (e.g., soil organic carbon, total nitrogen, rainfall, and air temperature). Understanding both biotic and abiotic factors which regulate soil biota diversity in tropical ecosystems and its role on ecosystem services represents the main challenge for both soil ecology and soil biology. We must consider that the main problem is the lack of information on the diversity of soil biota across spatiotemporal scale in tropical ecosystem (e.g., tropical rainforests, seasonal tropical forest, dry forests, spiny forests, desert, and islands). Despite this, we must try to understand and to develop theories on patterns of soil biota into tropical ecosystem, and the challenge is to identify the whole soil food web (e.g., here considering the spatiotemporal hierarchy of controls on soil biota diversity) and to determine the role of soil biodiversity as a driver of soil quality and environmental sustainability.

## References

Anyango JJ, Bautze D, Fiaboe KKM, Lagat ZO, Muriuki AW, Stöckli S, Riedel J, Onyambu GK, Musyoka MW, Karanja EN, Adamtey N (2020) The impact of conventional and organic farming on soil biodiversity conservation: a case study on termites in the long-term farming systems comparison trials in Kenya. BMC Ecol. https://doi.org/10.1186/s12898-020-00282-x

Bai P, Liu Q, Li X, Liu Y, Zhang L (2018) Response of the wheat rhizosphere soil nematode community in wheat/walnut intercropping system in Xinjiang, Northwest China. Appl Entomol Zool. https://doi.org/10.1007/s13355-018-0557-9

Balkenhol B, Haase H, Gebauer P, Lehmitz R (2018) Steeplebushes conquer the countryside: influence of invasive plant species on spider communities (Araneae) in former wet meadows. Biodivers Conserv 27:2257–2274. https://doi.org/10.1007/s10531-018-1536-8

Balota EL (2017) Manejo e qualidade biológica do solo. Editora Mecenas, Londrina

Bardgett RD (2005) The biology of soil: a community and ecosystem approach. Oxford University Press, New York

Basirat M, Mousavi SM, Abbaszadeh S, Ebrahimi M, Zarebanadkouki M (2019) The rhizosheath: a potential root trait helping plants to tolerate drought stress. Plant Soil 445:565–575. https://doi.org/10.1007/s11104-019-04334-0

Carrillo-Saucedo SM, Gavito ME, Siddique I (2018) Arbuscular mycorrhizal fungal spore communities of a tropical dry forest ecosystem show resilience to land-use change. Fungal Ecol. https://doi.org/10.1016/j.funeco.2017.11.006

Coyle DR, Nagendra UJ, Taylor MK, Campbell JH, Cunard CE, Joslin AH, Mundepi A, Phillips CA, Callaham MA Jr (2017) Soil fauna responses to natural disturbances, invasive species, and global climate change: current state of the science and a call to action. Soil Biol Biochem. https://doi.org/10.1016/j.soilbio.2017.03.008

de Novais CB, Sbrana C, Jesus EC, Rouws LFM, Giovannetti M, Avio L, Siqueira JO, Saggin Júnior OJ, da Silva EMR, de Faria SM (2020) Mycorrhizal networks facilitate the colonization of legume roots by a symbiotic nitrogen-fixing bacterium. Mycorrhiza. https://doi.org/10.1007/s00572-020-00948-w

Dial RJ, Ellwood MDF, Turner EC, Foster WA (2006) Arthropod abundance, canopy structure, and m microclimate in a Bornean lowland tropical rain forest. Biotropica 38(5):643–652. https://doi.org/10.1111/j.1744-7429.2006.00181.x

Doniger T, Adams JM, Marais E, Maggs-Kölling G (2020) The "fertile island effect" of *Welwitschia* plants on soil microbiota is influenced by plant gender. FEMS Microbiol. https://doi.org/10.1093/femsec/fiaa186

Forstall-Sosa KS, Souza TAF, Lucena EO, Silva SAI, Ferreira JTA, Silva TN, Santos D, Niemeyer JC (2020) Soil macroarthropod community and soil biological quality index in a green manure farming system of the Brazilian semi-arid. Biologia. In press. https://doi.org/10.2478/s11756-020-00602-y

Gebremikael MT, Steel H, Buchan D, Bert W, Neve S (2016) Nematodes enhance plant growth and nutrient uptake under C and N-rich conditions. Sci Rep. https://doi.org/10.1038/srep32862

Heydari M, Eslaminejad P, Kakhki FV, Mirab-Balou M, Omidipour R, Prévosto B, Kooch Y, Lucas-Borja MEL (2020) Soil quality and mesofauna diversity relationship are modulated by woody species and seasonality in semiarid oak forest. For Ecol Manag. https://doi.org/10.1016/j.foreco.2020.118332

Jo I, Friedley JD, Frank DA (2020) Rapid leaf litter decomposition of deciduous understory shrubs and lianas mediated by mesofauna. Plant Ecol. https://doi.org/10.1007/s11258-019-00992-3

Kamau S, Barrios E, Karanja NK, Ayuke FO, Lehmann J (2017) Soil macrofauna abundance under dominant tree species increases along a soil degradation gradient. Appl Soil Ecol. https://doi.org/10.1016/j.soilbio.2017.04.016

Kowalchuck GA, Stephen JR, de Boer W, Prosser JI, Embley TM, Woldendorp JW (1997) Analysis of ammonia-oxidising bacteria of the β-subdivision of the class *Proteobacteria* in coastal sand dunes by denaturing gradient gel electrophoresis and sequencing of PCR-amplified 16S ribosomal DNA fragments. Appl Environ Microbiol 63:1489–1497. https://doi.org/10.1128/aem.63.4.1489-1497.1997

Lammel DR, Cruz LM, Mescolotti D, Stürmer SL, Cardoso EJBN (2015) Woody Mimosa species are nodulated by Burkholderia in ombrophylous forest soils and their symbioses are enhanced by arbuscular mycorrhizal fungi (AMF). Plant Soil. https://doi.org/10.1007/s11104-015-2470-0

Lazzaro L, Mazza G, d'Errico G, Fabiani A, Giuliani IAF, Lagomarsino A, Landi S, Lastrucci L, Pastorelli R, Roversi PF, Torrini G, Tricarico E, Foggi B (2018) How ecosystems change following invasion by *Robinia pseudoacacia*: Insights from soil chemical properties and soil microbial, nematode, microarthropod and plant communities. Sci Total Environ 622-623:1509–1518. https://doi.org/10.1016/j.scitotenv.2017.10.017

Lira AFA, Vieira AGT, Olveira RF (2020) Seasonal influence on foraging activity of scorpion species (Arachnida: Scorpiones) in a seasonal tropical dry forest remnant in Brazil. Stud Neotropical Fauna Environ. https://doi.org/10.1080/01650521.2020.1724497

Liu H, Ding Y, Zhang Q, Liu X, Xu J, Li Y, Di H (2019a) Heterotrophic nitrification and denitrification are the main sources of nitrous oxide in two paddy soils. Plant Soil 445:39–53. https://doi.org/10.1007/s11104-018-3860-x

Liu H, Pan H, Hu H, Jia Z, Zhang Q, Liu Y, Xu J, Di H, Li Y (2019b) Autotrophic archaeal nitrification is preferentially stimulated by rice callus mineralization in a paddy soil. Plant Soil 445:55–69. https://doi.org/10.1007/s11104-019-04164-0

Lucena EO, Souza TAF, da Silva SIA, Kormann S, da Silva LJR, Laurindo LK, Forstall-Sosa KS, de Andrade LS (2021) Soil biota community composition as affected by *Cryptostegia madagascariensis* invasion in a tropical Cambisol from North-eastern Brazil. Trop Ecol. https://doi.org/10.1007/s42965-021-00177-y

Machado DL, Pereira MG, Correia MEF, Diniz AR, Menezes CEB (2015) Fauna edáfica na dinâmica sucessional da mata atlântica em floresta estacional semidecidual na bacia do rio Paraíba do Sul–RJ. Cienc Florest 25:91–106. https://doi.org/10.5902/1980509817466

Manwaring M, Wallace HM, Weaver HJ (2018) Effects of a mulch layer on the assemblage and abundance of mesostigmatan mites and other arthropods in the soil of a sugarcane agro-ecosystem in Australia. Exp Appl Acarol 74:291–300. https://doi.org/10.1007/s10493-0180227-1

Mauda EV, Joseph GS, Seymour CL, Munyai TC, Foord SH (2018) Changes in land use alter ant diversity, assemblage composition and dominant functional groups in African savannas. Biodivers Conserv 27:947–965. https://doi.org/10.1007/s10531-017-1474-x

Melo LN, Souza TAF, Santos D (2019) Cover crop farming system affects macroarthropods community diversity in Regosol of Caatinga, Brazil. Biologia. https://doi.org/10.2478/s11756-019-00272-5

Mirsam H, Muis A, Nonci N (2020) The density and diversity of plant-parasitic nematodes associated with maize rhizosphere in Malakaji Highland, South Sulawesi, Indonesia. Biodiversitas 21:2654–2661. https://doi.org/10.13057/biodiv/d210637

Myer A, Forschler BT (2019) Evidence for the role of subterranean termites (*Reticulitermes* spp.) in temperate forest soil nutrient cycling. Ecosystems 22:602–618. https://doi.org/10.1007/s10021-108-0291-8

Ng K, McIntyre S, Macfadyen S, Barton PS, Driscoll DA, Lindenmayer DB (2018) Dynamic effects of ground-layer plant communities on beetles in a fragmented farming landscape. Biodivers Conserv 27:2131–2153. https://doi.org/10.1007/s10531-018-1526-x

Nsengimana V, Kaplin AB, Nsabimana D, Dekoninck W, Francis F (2021) Diversity and abundance of soil-litter arthropods and their relationships with soil physicochemical properties under different land uses in Rwanda. Biodiversity 22(1-2):41–52. https://doi.org/10.1080/14888386.2021.1905064

Paymaneh Z, Sarcheshmehpour M, Bukovská P, Jansa J (2019) Could indigenous arbuscular mycorrhizal communities be used to improve tolerance of pistachio to salinity and/or drought? Symbiosis 79:269–283. https://doi.org/10.1007/s13199-019-00645-z

Raiesi F, Salek-Gilani S (2020) Development of a soil quality index for characterizing effects of land-use changes on degradation and ecological restoration of rangeland soils in a semi-arid ecosystem. Land Degrad Dev. https://doi.org/10.1002/ldr.3553

Rasmussen PU, Bennett AE, Tack AJM (2019) The impact of elevated temperature and drought on the ecology and evolution of plant-soil microbe interactions. J Ecol. https://doi.org/10.1111/1365-2745.13292

Real-Santillán RO, del-Val E, Cruz-Ortega R, Contreras-Cornejo HÁ, González-Esquivel CE, Larsen J (2019) Increased maize growth and P uptake promoted by arbuscular mycorrhizal fungi coincide with higher foliar herbivory and larval biomass of the fall armyworm *Spodoptera frugiperda*. Mycorrhiza 29:615–622. https://doi.org/10.1007/s00572-019-00920-3

Rodríguez J, Thompson V, Rubido-Bará M, Cordero-Rivera A, González L (2019) Herbivore accumulation on Invasive alien plants increases the distribution range of generalist herbivorous insects and supports proliferation of non-native insect pests. Biol Invasions 21:1511–1527. https://doi.org/10.1007/s10530-019-01913-1

Santos-Heredia C, Andresen E, Zárate DA, Escobar F (2018) Dung beetles and their Ecological functions in three agroforestry systems in the Lacandona rainforest of Mexico. Biodivers Conserv 27:2379–2394. https://doi.org/10.1007/s10531-018-1542-x

Sato T, Hachiya S, Inamura N, Ezawa T, Cheng W, Tawaraya K (2019) Secretion of acid phosphatase from extraradical hyphae of the arbuscular mycorrhizal fungus *Rhizophagus clarus* is regulated in response to phosphate availability. Mycorrhiza 29:599–605. https://doi.org/10.1007/s00572-019-00923-0

Sechi V, de Goede RGM, Rutgers M, Brussaard L, Mulder C (2018) Functional diversity in nematode communities across terrestrial ecosystems. Appl Soil Ecol 30:76–86. https://doi.org/10.1016/j.baae.2018.05.004

Silva RA, Siqueira GM (2020) Multifractal analysis of soil fauna diversity. Bragantia Indexes. https://doi.org/10.1590/1678-4499.20190179

Sousa NMF, Veresoglou SD, Oehl F, Rillig MC, Maia LC (2018) Predictors of arbuscular mycorrhizal fungal communities in the Brazilian Tropical Dry Forest. Soil Microbiol 75:447–458. https://doi.org/10.1007/s00248-017-1042-7

Souza TAF, Freitas H (2018) Long-Term effects of fertilization on soil organism diversity. In: Gaba S, Smith B, Lichtfouse E (eds) Sustainable agriculture reviews 28. Sustainable agriculture reviews. Springer, Cham. https://doi.org/10.1007/978-3-319-90309-57

Souza TAF, Rodrígues AF, Marques LF (2016) Long-term effects of alternative and conventional fertilization on macroarthropod community composition: a field study with wheat (*Triticum aestivum* L) cultivated on a Ferralsol. Org Agric 6:323–330. https://doi.org/10.1007/s13165-015-0138-y

Souza TAF, Santos (2018) Biologia dos Solos da Caatinga. Universidade Federal da Paraíba, PPGCS, Areia

Suleiman MK, Dixon K, Commander L, Nevill P, Quoreshi AM, Bhat NR, Manuvel A, Sivadasan MT (2019) Assessment of the Diversity of fungal community composition associated with *Vachellia pachyceras* and its rhizosphere soil from Kuwait desert. Front Microbiol. https://doi.org/10.3389/fmicb.2019.00063

Sun Y, Luo C, Jiang L, Song M, Zhang D, Li J, Li Y, Ostle NJ, Zhang G (2020) Land-use changes alter soil bacterial composition and diversity in tropical forest soil in China. Sci Total Environ. https://doi.org/10.1016/j.scitotenv.2020.136526

Swift MJ, Heal OW, Anderson JM (1979) Decomposition in terrestrial ecosystems. University of California Press, Berkeley

Tajik S, Ayiubi S, Lorenz N (2020) Soil microbial communities affected by vegetation, topography and soil properties in a forest ecosystem. Appl Soil Ecol 149:103514. https://doi.org/10.1016/j.apsoil.2020.103514

Tunlid A, White DC (1992) Biochemical analysis of biomass, community structure, nutritional status, and metabolic activity of microbial communities in soil. In: Stotzky G, Bollag JM (eds) Soil Biochemistry 7. Marcel Dekker, New York

Woese CR (1987) Bacterial evolution. Microbiol Rev 51:221–271

Yang B, Zhang W, Xu H, Wang S, Xu X, Fan H, Cheng HYH, Ruan H (2018) Effects of soil fauna on leaf litter decomposition under different land uses in eastern coast of China. J For Res 29(4):973–982. https://doi.org/10.1007/s11676-017-0521-5

# Chapter 3
# Soil Organisms and Ecological Processes

**Abstract** Previously, the soil ecosystem acting as habitat and food resource for soil organisms and the living soil and its wide variety of groups were introduced. This chapter will introduce the functional role of soil organisms on significant ecological processes, such as soil nutrient availability, biological control, soil structure, herbivory, symbiosis, and plant growth. Ecosystem engineers, litter transformers, predators, herbivores, symbionts, microregulators, decomposers, and prokaryotic transformers are essential for the soil ecosystem functioning because they promote services that provide habitat and food resource for the entire soil food web (e.g., including the primary producers). Basically, soil organisms are living individuals which live into soil ecosystem (e.g., by creating nests and galleries), and they have key role in increasing nutrient cycling, plant growth, soil structure, and plant resistance to abiotic and biotic stresses in tropical ecosystems. They can increase plant growth (e.g., by symbiotic relationships helping plants to uptake N, P, and micronutrients), litter deposition (e.g., by increasing the production of phytohormones to start leaf senescence processes), nutrient cycling (e.g., by increasing litter transformation and soil organic matter decomposition), and soil food web (e.g., by promoting a wide range of functional groups, and controlling the dominance of potential pathogen groups).

**Keywords** Biological control promoted by soil organisms · Plant growth · Symbionts in tropical ecosystems · Soil nutrient availability · Structuring soil profiles

**Questions Covered in the Chapter**
1. How do soil organisms affect important ecological processes into soil ecosystem?
2. How do litter transformers and decomposers promote soil nutrient availability?
3. How do microregulators and predators promote biological control?
4. How do ecosystem engineers promote soil structure?
5. How do symbionts help plants to uptake water and nutrients?
6. How does plant growth could be promoted by soil organism activity?

## 3.1  Introduction

Soil organisms (hereafter soil biota) are living individuals for a wide range of groups that are present into soil profile, and in tropical ecosystem, they play a role in providing important ecosystem services (Li et al. 2021). They can do the following:

1. Transform and incorporate the litter and other organic compounds that were deposited on soil surface thus having an important role in nutrient cycling.
2. Control population picks into their own community thus avoiding overpressure.
3. Make complex structures inside soil profile thus facilitating water movement and root growth.
4. Produce volatile compound that attracts herbivores and predators thus controlling leaf transpiration, photosynthetic rate, primary production, and herbivory pressure. At the tropics, in natural semiarid ecosystem, this process plays a key role on environmental sustainability.
5. Colonize plant roots thus promoting water and nutrient uptake.
6. Promote plant growth through biotic stresses thus stimulating phytohormone production, such as plant defensive response against herbivore insects.

Overall, soil organisms control litter deposition rate, soil organic matter transformation, plant mineral nutrition, and biological control (Andrea et al. 2018; Rodriguez et al. 2021;), increasing mineral nutrient availability (e.g., N, P, K, and micronutrients), plant growth (e.g., above- and belowground biomass production), and resistance tolerance to abiotic and biotic stresses in tropical conditions (Majeed et al. 2018; Yang et al. 2018). We can find soil organisms acting in several habitats into tropical ecosystems, such as forests, grasslands, shrublands, scrublands, and highly anthropogenized habitats. As they consist of a wide range of groups varying in their behavior, habitats, and food preferences, they exhibit different community composition, and it is dependent of plant community diversity, soil contents of C and N, and climate attributes (Zhang et al. 2017; Sprunger et al. 2019; Heydari et al. 2020). Some soil organism groups are dominant into soil profile (ecosystem engineers), like Blattidae, Formicidae, and Termitidae (Tuma et al. 2019a; Franco et al. 2020; Lavelle et al. 2020), but there are fewer dominant groups that play important functions (litter transformers and predators), like Scarabaeidae, Scutigeromorpha, and some soil nematodes (Liu et al. 2019; Gonçalves et al. 2021). Many works have also suggested that soil organisms may exhibit different distribution patterns and functional redundancy resulting in a high variability of diversity and composition between land uses or natural ecosystems (Silva et al. 2017; Potatov et al. 2019; Kooch et al. 2020).

## 3.2   Soil Nutrient Availability

Soil organisms are strongly dependent to C and N sources on soil profile, being unable to act improving soil ecosystem if these two nutrients were unavailable or in low contents (Potatov et al. 2017; Lin et al. 2017). Usually, in tropical ecosystems with high provision of habitat and food resources, soil organisms improve soil nutrient availability following three distinct pathways (Fig. 3.1).

They start degrading whole litter material by physical trituration (Joly et al. 2020; Brosseau et al. 2019) through their mouthparts (e.g., litter transformers are characterized by presenting a robust pair of mandibles). Next, they carry litter residues into soil profile by bioturbation process (Sofo et al. 2020a; McCary and Schimitz 2021). Here, litter transformers promote soil excavation, seed exhumation (e.g., by improving soil seed bank viability), and L-layer formation and provide habitat to generalist decomposers (e.g., yeast) which in turn start degrading litter residues by enzymatic activity (Urrea-Galeano et al. 2019; Lucena et al. 2021). One important point during this process is that we can split soil mineral phase from litter residues. Finally, once the soil organic matter has been produced, only specialist decomposers act promoting decomposition (e.g., mineralization and humification) and nutrient cycling process (López-Mondéjar et al. 2018; Brunel et al. 2020; Husson et al. 2021). This final pathway occurs exclusively through chemical alteration promoted by microbial enzymes (Table 3.1).

## 3.3   Herbivory

Probably, the interaction between plant and soil organisms classified as herbivores (e.g., arthropods and nematodes) is the most ancient ecological interaction in the humankind history (Currano et al. 2021). During this process, the soil organisms feed several parts of the plant (e.g., leaves, roots, branches, flowers, fruits, and seeds) thus releasing chemical signals that induce plants to start their defense response (Yang et al. 2021). Here, it is important to know that the herbivory is an important and natural process that occurs all the time around all the world. Into this context, it is incorrect to classify a herbivore as a "pest" (Heinze 2020). In natural conditions where there is a complex habitat with a high diversity of plant species, the herbivory is beneficial for the plant species (Cortois et al. 2017; Wan et al. 2020b). It helps the plant species into four different ways:

1. It reduces plant transpiratory rate by reducing the leaf area. It is so necessary into semiarid ecoregions where there is a long dry period.
2. It acts as an "enzyme" catalyzing the leaf senescence stage thus promoting the litter deposition.
3. It enables the plant species to produce phytohormones and pigments that attract predators thus promoting biological control.
4. Leaf-feeders can promote pollination by transporting pollen.

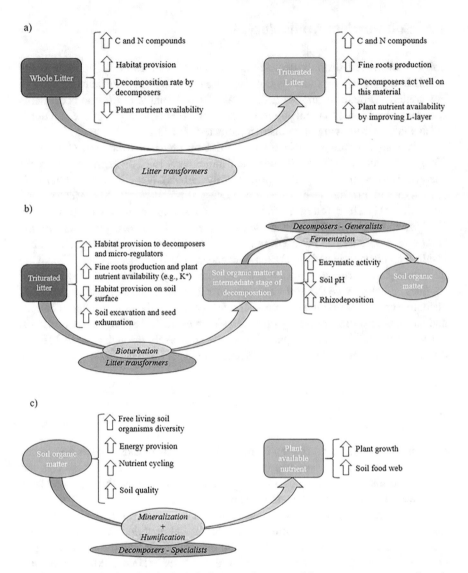

**Fig. 3.1** Schematic view showing how do litter transformers and decomposers promote soil nutrient availability in tropical ecosystems: (**a**) physical pathway (litter trituration), (**b**) physical-chemical pathway, and (**c**) chemical pathway (enzymes). (Adapted from Wutzler et al. (2017), Magcale-Macandog et al. (2018), Griffiths et al. (2021), Homet et al. (2021))

However, in tropical agriculture where the natural ecosystem is simplified into monocropping systems, the herbivore pressure on annual plant species is too high because of the reduction of plant diversity, habitat provision, and food availability. Thus, creating a negative feedback and a hard scenario where the herbivores impose

**Table 3.1** Main enzymes produced by decomposers

| Enzyme | Action |
|---|---|
| Amidase | It catalyzes the breakdown of N-rich organic compounds into ammonium $NH_4$ |
| Amylase | It catalyzes the breakdown of starch into sugar |
| Aminoacylase | It catalyzes the N-acyl-L-amino acid into carboxylate and L-amino acid |
| Aminopeptidase | It catalyzes the cleavage of amino acids into proteins or exopeptidases |
| Asparaginase | It catalyzes the breakdown of asparagine into ammonia and aspartic acid |
| Catalase | It catalyzes the hydrogen peroxide into water and oxygen |
| Cellulase | It catalyzes the cellulose molecule into monosaccharides |
| Cutinase | It catalyzes the breakdown of cutin into cutin monomers |
| FDA hydrolysis | It catalyzes the decomposition of soil organic matter into glucose |
| Galactosidase | It catalyzes the hydrolysis of galactosides into monosaccharides |
| Glucosidase | It catalyzes the breakdown of carbohydrates into glucose |
| Glucose isomerase | It catalyzes the interconversion of D-xylose and D-xylulose |
| Glucose oxidase | It catalyzes the glucose into hydrogen peroxide and D-glucono-δ-lactone |
| Glucoamylase | It catalyzes the alpha-D-glucose residues into beta-D-glucose |
| Invertase | It catalyzes the breakdown of sucrose into fructose and glucose |
| Laccase | It oxidizes a variety of phenolic substrates, such as lignin |
| Lactase | It catalyzes hydrolysis of the lactose into galactose and glucose monomers |
| Lipase | It catalyzes the breakdown or hydrolysis of fats |
| Oxygenase | It oxidizes a substrate by transferring the oxygen from $O_2$ to it |
| Pectinase | It catalyzes the breakdown of pectin |
| Phosphatase | It catalyzes the breakdown of inorganic P into phosphate |
| Protease/ Proteinase | It catalyzes the breakdown of proteins into smaller polypeptides |
| Phytase | It catalyzes the hydrolysis of phytic acid into inorganic phosphorus |
| Sulfatase | It catalyzes the breakdown of sulfur into sulfate |
| Urease | It catalyzes the breakdown of urea into ammonia and carbon dioxide |
| Transglutaminase | It catalyzes the formation of an isopeptide bond between glutamine residue and lysine residue with subsequent release of ammonia |
| Tyrosinase | It catalyzes the oxidation of phenols such as tyrosine and dopamine using dioxygen |
| Xylanase | It catalyzes the breakdown of linear polysaccharide xylan into xylose |

Adapted from Ba and Kumar (2017), Garske et al. (2017), Ren et al. (2017), Dotaniya et al. (2019); Gunjal et al. (2019), Lincoln et al. (2019), Sorde and Ananthanarayan (2019), Shi et al. (2019), Unuofin et al. (2019), Hou et al. (2020), Norman et al. (2020), Vara and Karnena (2020), Wan et al. (2020a), Wilkerson and Olapade (2020), Ivaldi et al. (2021)

a stress that annual plant species have no conditions to resist by their own (Bennett and Klironomos 2018). In general, we can find two situations regarding the herbivory into the tropics: a positive and a negative feedback related to natural ecosystem and monocropping system, respectively (Fig. 3.2).

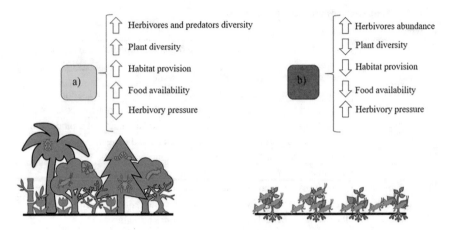

**Fig. 3.2** Schematic view showing how do herbivores and plant species interact in (**a**) natural ecosystem and (**b**) monocropping system. (Adapted from Yang et al. (2018), Bennett et al. (2019), Wan et al. (2019))

## 3.4   Biological Control

It is another natural ecological process that occurs into soil ecosystem constantly, but it is extremely sensitive to changes and disturbances both at temporal and spatial scale (Neher and Barbercheck 2019). We can define the biological control into the tropical soils as the sum of predators and microregulators activity (Marsden et al. 2020). Here, predators (e.g., Chilopoda, Dermaptera, Mantodea, Scorpiones, and some families from Coleoptera) act as controlling agent in the population of individuals from macrofauna and mesofauna groups, especially herbivores, while microregulators (e.g., Acari, Collembola, some carnivores Nematodes, and other microorganisms) act as controlling agents in microbiota population into soil profile and around the rhizosphere microbiome (Lazarova et al. 2021). Because of this process in natural ecosystem, we never find high herbivory pressure or high abundance of a specific group (e.g., high abundance of larvae of Coleoptera) at the tropics (Ciancio and Gamboni 2017; Sankaranarayanan et al. 2019). Here, it is important to clarify the difference between the idea of dominance and abundance into soil ecology (Arnan et al. 2018). Even in natural ecosystem, we can naturally find dominant groups such as ecosystem engineers (e.g., Formicidae and Termitidae), and it does not reflect a disturbance or a negative feedback (Tuma et al. 2019b; Vanolli et al. 2021). It is a natural phenomenon following the biology and behavior of ants and termites to live into colonies (Menzel and Feldmeyer 2021). On the other hand, abundance reflects the number of a specific taxon for an area or a specific ecosystem (de Cristo et al. 2019). High abundance rates for a specific taxon, i.e., herbivores from Cerambycidae and Hemiptera and larvae of Lepidoptera, Orthoptera, or Phasmatodea, could reflect disturbance, reduction into plant diversity, and low predator abundance. In this context, we can find a reduction on the biological control, such as in the case of tropical monocropping systems where there is a high herbivory pressure promoted by the reduction of food availability and habitat provision that in turn leads to a drastic reduction on predator abundance (Dietrich et al. 2021).

## 3.5  Soil Structure

Ecosystem engineers (e.g., ants, cockroaches, earthworms, isopods, and termites) affect soil structure by promoting bioturbation and physical alteration of the organic matter (e.g., consumption of litter and production of fecal pellets) and building physical structures in soil profile that act as habitat for decomposers and prokaryotic transformers (Piron et al. 2017; Jouquet et al. 2018; Domínguez et al. 2018; Sagi and Hawlena 2021). During bioturbation, soil organisms promote the movement of materials through soil surface and soil profile. Litter transformers promoted soil excavation, litter incorporation (e.g., contributing to O horizon formation), seeds dispersion and exhumation (McTavish and Murphy 2021; Wiesmeier et al. 2019). In soil profile, they consume organic detritus and produce fecal pellets. These pellets provide a favorable habitat to decomposers, thus increasing decomposition rate and nutrient release (Sagi et al. 2021). Finally, soil engineers create a complex chain of channels because of their activity. It positively modifies the soil macropores (e.g., also classified as biopores), soil porosity, and drainage (Lehmann et al. 2017). One important point into this process is that all activities are exclusively promoted by soil organisms from macrofauna group (Gongalsky 2021).

## 3.6  Symbiosis

Some individuals in Fungi and Bacteria groups (e.g., arbuscular mycorrhizal fungi and N-fixing bacteria) can establish a symbiotic relationship with plant species (Porter and Sachs 2020). These microorganisms are present in all tropical ecosystems and have an important role in both plant mineral nutrition (e.g., N, P, and micronutrient uptake) and water absorption (Winagraski et al. 2019; Medeiros et al. 2021). They produce structures inside roots (e.g., arbuscules for fungi from Phylum Glomeromycota and nodules for bacteria from Family Rhizobiaceae) (Uroz et al. 2019) that act as a specialized tool for the exchange of nutrients between the symbionts (Jacott et al. 2017). The AMF symbiosis results in increased plant growth, resistance, and tolerance to abiotic and biotic stresses (Naik et al. 2019). After primary production (Fig. 3.3), the symbiosis is the second basis of the soil food web (Antunes and Koyama 2017).

  Into Glomeromycota, the arbuscules are originated among the cell wall and plasma membrane of root cortical cells by differentiating the intracellular hyphae (Souza 2015), while into Rhizobiaceae, the nodules are originated by creating a pathway for the bacterial cell to travel inside the root epidermal cells

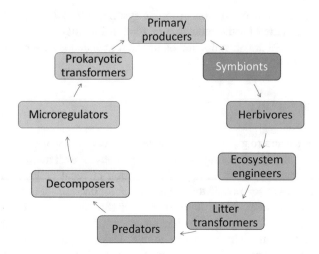

**Fig. 3.3** Schematic view of the entire soil food web. (Adapted from Veldhuis et al. (2018))

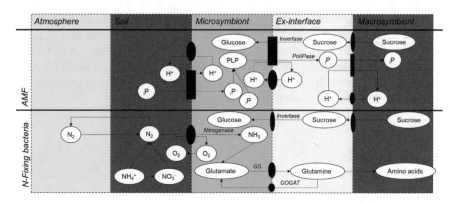

**Fig. 3.4** Main events of the symbiotic process following the pathways for an arbuscular mycorrhizal fungus (AMF) and a nitrogen fixing (N-fixing) bacteria. (Adapted from Chang et al. (2017), Meng et al. (2018))

(Gaudioso-Pedraza et al. 2018). The nodulation begins when the nod factor initiates root hair curling around the bacterial cell (Zanetti et al. 2020). Here, we must consider that the symbiotic relationship is modulated by specific genes thus creating different levels of host specificity during symbiont pairing (Andrews et al. 2018; Kimeklis et al. 2019). Also, we can find some plant species that present inside their roots both fungi and bacteria acting as symbionts. This phenomenon is called tripartite symbiosis and occurs most frequently in legumes (Wang et al. 2021).

In general, the main function of the symbiosis is to provide for the host-plant essential nutrients through its colonized roots (Ingraffia et al. 2019). Even the function being the same, fungi and bacteria (e.g., arbuscular mycorrhizal fungi and N-fixing bacteria) exchange nutrients in totally different pathways (Fig. 3.4).

Arbuscular mycorrhizal fungi provide nutrients, i.e., in general inorganic P ($_iP$), but they can also provide other macro- and micronutrients, for the host plant, which in turn provide to the fungus C-rich compounds (e.g., glucose) (Lanfranco et al. 2019). Here the nutrient exchange is modulated by H+ extrusion at soil-microsymbiont, microsymbiont-exchange-interface, and exchange-interface-macrosymbiont zones (da Silva et al. 2021), thus being extremely sensitive to changes in soil pH (See $H^+$ extrusion hypothesis proposed by Ramos et al. 2009) and dependent to a specific enzyme (e.g., PoliPase) that breaks down the polyphosphate molecule into single iP at exchange interface zone (Lorenzo-Orts et al. 2019; Recorbet et al. 2021). For the N-fixing bacteria, the symbiotic process occurs totally different from the AMF pathway (Foo 2019; Sakamoto et al. 2019). Here, the bacteria need an aerobic condition (e.g., the presence of dissolved $O_2$ into soil solution) (García-Cabrera et al. 2020; Rutten and Poole 2019; Verma et al. 2020). It activates the nitrogenase enzyme at the microsymbiont structure (e.g., root nodules), thus being extremely sensitive to changes in soil aeration and soil water regime (Nishida and Suzaki 2018; Schwember et al. 2019; Rutten et al. 2021). Other three enzymes work into this pathway, i.e., invertase by converting sucrose into glucose, *GS* by converting glutamate into glutamine, and finally GOGAT by converting glutamine into glutamate, thus controlling the biological fixation of N and plant uptake (Santos et al. 2018; Fonseca-García et al. 2021).

## 3.7  Plant Growth

As already shown, all the previous processes can directly and indirectly promote plant growth (Souza and Santos 2018; Lekberg et al. 2018). However, our focus here will be on mineralization that is governed by decomposers and prokaryotic transformers (Hellequin et al. 2018; Suleiman et al. 2019). In general, there are three main pathways that both decomposers and prokaryotic transformers can affect nutrient mineralization:

1. Selective feeding on microbiota: Promoted by selective soil organisms (e.g., fungivores and bacterivores) that choose to feed particular individuals from microbiota group over others. It can lead to positive changes in soil-plant feedback and has a dramatic effect on prey's activity as described by the *"knock-on"* (Iordache 2020).
2. Altering the soil organic matter forms (e.g., litter, F, and H layers): microbial-feeding fauna enhances C and N mineralization that resulted from increased turnover rate and microbial activity (e.g., respiration) and from the excretion of excessive N on microbial cell, respectively (Thakur and Geisen 2019; Sofo et al. 2020b).
3. Hormonal effects on root morphology: Also described as a non-nutritional effect, it is played by protozoa, some nematodes, and other bacteria taxa that influence plant species to produce growth-promoting hormones (e.g., auxins) through their

activity near the rhizosphere (Keswani et al. 2020) thus promoting changes in root architecture (e.g., plant species developing a highly branched root system) (Bavaresco et al. 2020; Pan et al. 2020; Lynch et al. 2021).

## 3.8  Conclusions

Soil organisms promote important service in the ecosystem processes of soil nutrient availability, biological control, soil structure, herbivory, symbiosis, and plant growth. They are essential for the soil ecosystem functioning because they promote services that provide habitat and food resource for the entire soil food web (e.g., including the primary producers). They promote plant growth (e.g., by symbiotic relationships helping plants to uptake N, P, and micronutrients), litter deposition (e.g., by increasing the production of phytohormones to start leaf senescence processes), nutrient cycling (e.g., by increasing litter transformation and soil organic matter decomposition), and soil food web (e.g., by promoting a wide range of functional groups and controlling the dominance of potential pathogen groups). However, even knowing all these processes, we must consider that they occur simultaneously interacting with each other that has cascading effects on soil food web. As already described, soil organisms can influence soil structure, nutrient availability, and primary production (e.g., measured by the plant growth), thereby affecting the habitat provision and the food availability to the entire soil food web.

## References

Andrea F, Bini C, Amaducci S (2018) Soil and ecosystem services: current knowledge and evidences from Italian case studies. Appl Soil Ecol 123:693–698. https://doi.org/10.1016/j.apsoil.2017.06.031

Andrews M, De Meyer S, James EK, Stępkowski T, Hodge S, Simon MF, Young JPW (2018) Horizontal transfer of symbiosis genes within and between Rhizobial genera: occurrence and importance. Genes 9(7):321. https://doi.org/10.3390/genes9070321

Antunes PM, Koyama AA (2017) Mycorrhizas as nutrient and energy pumps of soil food webs: multitrophic interactions and feedbacks. In: Johnson NC, Gehring C, Jansa J (eds) Mycorrhizal mediation of soil fertility, structure, and carbon storage. Elsevier. https://doi.org/10.1016/B978-0-12-804312-7.00009-7

Arnan X, Andersen AN, Gibb H, Parr CL, Sanders NJ, Dunn RR, Angulo E, Baccaro FB, Bishop TR, Boulay R, Castracani C, Cerdá X, Del Toro I, Delsinne T, Donoso DA, Elten EK, Fayle TM, Fitzpatrick MC, Gómez C, Grasso DA, Grossman BF, Guénard B, Gunawardene N, Heterick B, Hoffmann BD, Janda M, Jenkins CN, Klimes P, Lach L, Laeger T, Leponce M, Lucky A, Majer J, Menke S, Mezger D, Mori A, Moses J, Munyai TC, Paknia O, Pfeiffer M, Philpott SM, Souza JLP, Tista M, Vasconcelos HL, Retana J (2018) Dominance-diversity relationships in ant communities differ with invasion. Glob Chang Biol 24(10):4614–4625. https://doi.org/10.1111/gcb.14331

Ba S, Kumar VV (2017) Recent developments in the use of tyrosinase and laccase in environmental applications. Crit Rev Biotechnol 37(7):819–832. https://doi.org/10.1080/0738855 1.2016.1261081

Bavaresco LG, Osco LP, Araujo ASF, Mendes LW, Bonifacio A, Araújo FF (2020) *Bacillus subtilis* can modulate the growth and root architecture in soybean through volatile organic compounds. Theor Exp Plant Physiol 32:99–108. https://doi.org/10.1007/s40626-020-00173-y

Bennett JA, Klironomos J (2018) Mechanisms of plant–soil feedback: interactions among biotic and abiotic drivers. New Phytol 222:91–96. https://doi.org/10.1111/nph.15603

Bennett JA, Koch AM, Forsythe J, Johnson NC, Tilman D, Klironomos J (2019) Resistance of soil biota and plant growth to disturbance increases with plant diversity. Ecol Lett 23:119–128. https://doi.org/10.1111/ele.13408

Brosseau P, Gravel D, Handa IT (2019) Traits of litter-dwelling forest arthropod predators and detritivores covary spatially with traits of their resources. Ecology 100(10):e02815. https://doi.org/10.1002/ecy.2815

Brunel C, da Silva AMF, Gros R (2020) Environmental drivers of microbial functioning in mediterranean forest soils. Microb Ecol 80:669–681. https://doi.org/10.1007/s00248-020-01518-5

Chang C, Nasir F, Ma L, Tian C (2017) Molecular communication and nutrient transfer of arbuscular mycorrhizal fungi, symbiotic nitrogen-fixing bacteria, and host plant in Tripartite Symbiosis. In: Sulieman S, Tran LS (eds) Legume nitrogen fixation in soils with low phosphorus availability. Springer, Cham. https://doi.org/10.1007/978-3-319-55729-8_9

Ciancio A, Gamboni M (2017) Soil biodiversity and tree crops resilience. In: Lukac M, Grenni P, Gamboni M (eds) Soil biological communities and ecosystem resilience. Sustainability in plant and crop protection. Springer, Cham. https://doi.org/10.1007/978-3-319-63336-7_20

Cortois R, Veen GFC, Duyts H, Abbas M, Strecker T, Kostenko O, Eisenhauer N, Scheu S, Gleixner G, de Deyn GB, van der Putten WH (2017) Possible mechanisms underlying abundance and diversity responses of nematode communities to plant diversity. Ecosphere 8(5):e01719. https://doi.org/10.1002/ecs2.1719

Currano ED, Azevedo-Schmidt LE, Maccracken AS, Swain A (2021) Scars on fossil leaves: an exploration of ecological patterns in plant–insect herbivore associations during the age of angiosperms. Paleogeogr Palaeoclimatol Palaeoecol 582:110636. https://doi.org/10.1016/j.palaeo.2021.110636

da Silva SIA, Souza TAF, Lucena EO, da Silva LJR, Laurindo LK, Nascimento GS, Santos D (2021) High phosphorus availability promotes the diversity of arbuscular mycorrhizal spores' community in different tropical crop systems. Biologia. https://doi.org/10.1007/s11756-021-00874-y

de Cristo SC, Vitorino MD, Arenhardt TCP, Klunk GA, Adenesky Filho E, de Carvalho AG (2019) Leaf-litter Entomofauna as a parameter to evaluate areas under ecological restoration. Floresta e Ambiente 26(2):e20170295. https://doi.org/10.1590/2179-8087.029517

Dietrich P, Cesarz S, Liu T, Roscher C, Eisennahuer N (2021) Effects of plant species diversity on nematode community composition and diversity in a long-term biodiversity experiment. Oecologia. https://doi.org/10.1007/s00442-021-04956-1

Domínguez A, Jiménez JJ, Ortíz CE, Bedano JC (2018) Soil macrofauna diversity as a key element for building sustainable agriculture in argentine pampas. Acta Oecol 92:102–116. https://doi.org/10.1016/j.actao.2018.08.012

Dotaniya ML, Aparna K, Dotaniya CK, Singh M, Regar KL (2019) Role of soil enzymes in sustainable crop production. In: Kuddus M (eds) Enzymes in food biotechnology, 569–589. https://doi.org/10.1016/B978-0-12-813280-7.00033-5

Fonseca-García C, Nava N, Lara M, Quinto C (2021) An NADPH oxidase regulates carbon metabolism and the cell cycle during root nodule symbiosis in common bean (*Phaseolus vulgaris*). BMC Plant Biol 21:274. https://doi.org/10.1186/s12870-021-03060-z

Foo, E. (2019). Plant hormones play common and divergent roles in nodulation and arbuscular mycorrhizal symbioses. In: de Bruijn F (eds) The Model Legume Medicago truncatula. Wiley. https://doi.org/10.1002/9781119409144.ch93

Franco ALC, Cherubin MR, Cerri CEP, Six J, Wall DH, Cerri CC (2020) Linking soil engineers, structural stability, and organic matter allocation to unravel soil carbon responses to land-use change. Soil Biol Biochem 150:107998. https://doi.org/10.1016/j.soilbio.2020.107998

García-Cabrera RI, Valdez-Cruz NA, Blancas-Cabrera A, Trujillo-Roldán MA (2020) Oxygen transfer rate affect polyhydroxybutyrate production and oxidative stress response in submerged cultures of *Rhizobium phaseoli*. Biochem Eng J 162:107721. https://doi.org/10.1016/j.bej.2020.107721

Garske AL, Kapp G, McAuliffe JC (2017) Industrial enzymes and biocatalysis. In: Kent J, Bommaraju T, Barnicki S (eds) Handbook of industrial chemistry and biotechnology. Springer, Cham. https://doi.org/10.1007/978-3-319-52287-6_28

Gaudioso-Pedraza R, Beck M, Frances L, Kirk P, Ripodas C, Niebel A, Oldroyd GED, Benitez-Alfoso Y, Carvalho-Niebel F (2018) Callose-regulated symplastic communication coordinates symbiotic root nodule development. Curr Biol 28(22):3562–3577.e6. https://doi.org/10.1016/j.cub.2018.09.031

Gonçalves F, Carlos C, Crespo L, Zina V, Oliveira A, Salvação J, Pereira JA, Torres L (2021) Soil arthropods in the Douro demarcated region vineyards: general characteristics and ecosystem services provided. Sustainability 13(14):7837. https://doi.org/10.3390/su13147837

Gongalsky KB (2021) Soil macrofauna: study problems and perspectives. Soil Biol Biochem 159:108281. https://doi.org/10.1016/j.soilbio.2021.108281

Griffiths HM, Ashton LA, Parr CL, Eggleton P (2021) The impact of invertebrate decomposers on plants and soil. New Phytol 231(6):2142–2149. https://doi.org/10.1111/nph.17553

Gunjal AB, Waghmode MS, Patil NN, Nawani NN (2019) Significance of soil enzymes in agriculture. In: Bhatt P (ed) Smart bioremediation technologies, 159–168. https://doi.org/10.1016/B978-0-12-818307-6.00009-3

Heinze J (2020) Herbivory by aboveground insects impacts plant root morphological traits. Plant Ecol 221:725–732. https://doi.org/10.1007/s11258-020-01045-w

Hellequin E, Monard C, Quaiser A, Henriot M, Klarzynski O, Binet F (2018) Specific recruitment of soil bacteria and fungi decomposers following a biostimulant application increased crop residues mineralization. PLoS One 13(12):e0209089. https://doi.org/10.1371/journal.pone.0209089

Heydari M, Eslaminejad P, Kakhki FV, Mirab-balou M, Omidipour R, Prévosto B, Koock Y, Lucas-Borja ME (2020) Soil quality and mesofauna diversity relationship are modulated by woody species and seasonality in semiarid oak forest. For Ecol Manag 473:118332. https://doi.org/10.1016/j.foreco.2020.118332

Homet P, Gómez-Aparicio L, Matías L, Godoy O (2021) Soil fauna modulates the effect of experimental drought on litter decomposition in forests invaded by an exotic pathogen. J Ecol 109(8):2963–2980. https://doi.org/10.1111/1365-2745.13711

Hou W, Wang J, Nan Z, Christensen MJ, Xia C, Chen T, Zhang Z, Niu X (2020) *Epichloë gansuensis* endophyte-infection alters soil enzymes activity and soil nutrients at different growth stages of *Achnatherum inebrians*. Plant Soil 455:227–240. https://doi.org/10.1007/s11104-020-04682-2

Husson O, Sarthou J, Ratnadass A, Schmidt H, Kempf J, Husson B, Tingry S, Aubertot J, Ddeguine J, Goebel F, Lamichhane JM (2021) Soil and plant health in relation to dynamic sustainment of Eh and pH homeostasis: a review. Plant Soil 466:391–447. https://doi.org/10.1007/s11104-021-05047-z

Iordache V (2020) On the possibility of accelerating succession by manipulating soil microorganisms. In: Varma A, Tripathi S, Prasad R (eds) Plant microbiome paradigm. Springer, Cham. https://doi.org/10.1007/978-3-030-50395-6_11

Ingraffia R, Amato G, Frenda AS, Giambalvo D (2019) Impacts of arbuscular mycorrhizal fungi on nutrient uptake, $N_2$ fixation, N transfer, and growth in a wheat/faba bean intercropping system. PLoS One 14(3):e0213672. https://doi.org/10.1371/journal.pone.0213672

Ivaldi C, Daou M, Vallon L, Bisotto A, Haon M, Garajova S, Bertrand E, Faulds CB, Sciara G, Jacotot A, Marchand C, Hugoni M, Rakotoarivonina H, Rosso M-N, Rémond C, Luis P, Record E (2021) Screening new xylanase biocatalysts from the mangrove soil diversity. Microorganisms 9(7):1484. https://doi.org/10.3390/microorganisms9071484

Jacott CN, Murray JD, Ridout CJ (2017) Trade-offs in arbuscular mycorrhizal symbiosis: disease resistance, growth responses and perspectives for crop breeding. Agronomy 7(4):75. https://doi.org/10.3390/agronomy7040075

Joly FX, Coq S, Coulis M, David J, Hätternschwiler S, Mueller CW, Subke J (2020) Detritivore conversion of litter into faeces accelerates organic matter turnover. Commun Biol 3:660. https://doi.org/10.1038/s42003-020-01392-4

Jouquet P, Chaudhary E, Kumar ARV (2018) Sustainable use of termite activity in agro-ecosystems with reference to earthworms. A review. Agron Sustain Dev 38:3. https://doi.org/10.1007/s13593-017-0483-1

Keswani C, Singh SP, Cueto L, García-Estrada C, Mezaache-Aichour S, Glare TR, Borriss R, Singh SP, Blázquez MA, Sansinenea E (2020) Auxins of microbial origin and their use in agriculture. Appl Microbiol Biotechnol 104:8549–8565. https://doi.org/10.1007/s00253-020-10890-8

Kimeklis AK, Chirak ER, Kuznetsova IG, Sazanova AL, Safronova VI, Belimov AA, Onishchuk OP, Kurchak ON, Aksenova TS, Pinaev AG, Andronov EE, Provorov NA (2019) Rhizobia isolated from the relict legume *Vavilovia formosa* represent a genetically specific group within *rhizobium leguminosarum* biovar *viciae*. Genes 10(12):991. https://doi.org/10.3390/genes10120991

Kooch Y, Ehsani S, Akabarinia M (2020) Stratification of soil organic matter and biota dynamics in natural and anthropogenic ecosystems. Soil Tillage Res 200:104621. https://doi.org/10.1016/j.still.2020.104621

Lanfranco L, Fiorilli V, Gutjahr C (2019) Partner communication and role of nutrients in the arbuscular mycorrhizal symbiosis. New Phytol 220(4):1031–1046. https://doi.org/10.1111/nph.15230

Lavelle P, Spain A, Fonte S, Bedano JC, Blanchart E, Galindo V, Grimaldi M, Jimenez JJ, Velasquez E, Zangerlé A (2020) Soil aggregation, ecosystem engineers and the C cycle. Acta Oecologia 105:103561. https://doi.org/10.1016/j.actao.2020.103561

Lazarova S, Coyne D, Rodríguez MG, Peteira B, Ciancio A (2021) Functional diversity of soil nematodes in relation to the impact of agriculture—a review. Diversity 13(2):64. https://doi.org/10.3390/d13020064

Lehmann A, Zheng W, Rillig MC (2017) Soil biota contributions to soil aggregation. Nat Ecol Evol 1:1828–1835. https://doi.org/10.1038/s41559-017-0344-y

Lekberg Y, Bever JD, Bunn RA, Callaway RM, Hart MM, Kivlin SN, Klironomos J, Larkin BG, Maron JL, Reinhart KO, Kemke M, Remke M, van der Putten WH (2018) Relative importance of competition and plant–soil feedback, their synergy, context dependency and implications for coexistence. Ecol Lett 21:1268–1281. https://doi.org/10.1111/ele.13093

Li K, Zhang H, Li X, Wang C, Zhang J, Jiang R, Feng G, Liu X, Zuo Y, Yuan H, Zhang C, Gai J, Tian J (2021) Field management practices drive ecosystem multifunctionality in a smallholder-dominated agricultural system. Agric Ecosyst Environ 313:107389. https://doi.org/10.1016/j.agee.2021.107389

Lin H, He Z, Hao J, Tian K, Jia X, Kong X, Akbar S, Bei Z, Tian X (2017) Effect of N addition on home-field advantage of litter decomposition in subtropical forests. For Ecol Manag 398:216–225. https://doi.org/10.1016/j.foreco.2017.05.015

Lincoln L, More VS, More SS (2019) Purification and biochemical characterization of extracellular glucoamylase from *Paenibacillus amylolyticus* strain. J Basic Microbiol 59:375–384. https://doi.org/10.1002/jobm.201800540

Liu T, Hu F, Li H (2019) Spatial ecology of soil nematodes: perspectives from global to micro scales. Soil Biol Biochem 137:107565. https://doi.org/10.1016/j.soilbio.2019.107565

López-Mondéjar R, Brabcová V, Štursová M, Davidová A, Jansa J, Cajthaml T, Baldrian P (2018) Decomposer food web in a deciduous forest shows high share of generalist microorganisms and importance of microbial biomass recycling. ISME J 12:1768–1778. https://doi.org/10.1038/s41396-018-0084-2

Lorenzo-Orts L, Couto D, Hothorn M (2019) Identity and functions of inorganic and inositol polyphosphates in plants. New Phytol 225(2):637–652. https://doi.org/10.1111/nph.16129

Lucena EO, Souza TAF, da Silva SIA, Kormann S, da Silva LJR, Laurindo LK, Forstall-Sosa KS, de Andrade LS (2021) Soil biota community composition as affected by *Cryptostegia madagascariensis* invasion in a tropical Cambisol from North-Eastern Brazil. Trop Ecol. https://doi.org/10.1007/s42965-021-00177-y

Lynch JP, Strock CF, Schneider HM, Sidhu JS, Ajmera I, Galino-Casteñada T, Klein SP, Hanlon MT (2021) Root anatomy and soil resource capture. Plant Soil 466:21–63. https://doi.org/10.1007/s11104-021-05010-y

Magcale-Macandog DB, Manlubatan MBT, Javier JM, Edrial JD, Mago KS, de Luna JEI, Nayoos J, Porcioncula RP (2018) Leaf litter decomposition and diversity of arthropod decomposers in tropical *Muyong* forest in Banaue, Philippines. Paddy Water Environ 16:265–277. https://doi.org/10.1007/s10333-017-0624-9

Majeed A, Muhammad Z, Ahmad H (2018) Plant growth promoting bacteria: role in soil improvement, abiotic and biotic stress management of crops. Plant Cell Rep 37:1599–1609. https://doi.org/10.1007/s00299-018-2341-2

Marsden C, Martin-Chave A, Cortet J, Hedde M, Capowiez Y (2020) How agroforestry systems influence soil fauna and their functions – a review. Plant Soil 453:29–44. https://doi.org/10.1007/s11104-019-04322-4

McTavish MJ, Murphy SD (2021) Three-dimensional mapping of earthworm (*Lumbricus terrestris*) seed transport. Pedobiologia 87-88:150752. https://doi.org/10.1016/j.pedobi.2021.150752

McCary MA, Schimitz OJ (2021) Invertebrate functional traits and terrestrial nutrient cycling: insights from a global meta-analysis. J Anim Ecol 90(7):1714–1726. https://doi.org/10.1111/1365-2656.13489

Medeiros AS, Goto BT, Ganade G (2021) Ecological restoration methods influence the structure of arbuscular mycorrhizal fungal communities in degraded drylands. Pedobiologia 84:150690. https://doi.org/10.1016/j.pedobi.2020.150690

Meng S, Wang S, Quan J, Su W, Lian C, Wang D, Xia X, Yin W (2018) Distinct carbon and nitrogen metabolism of two contrasting poplar species in response to different N supply levels. Int J Mol Sci 19(8):2302. https://doi.org/10.3390/ijms19082302

Menzel F, Feldmeyer B (2021) How does climate change affect social insects? Curr Opin Insect Sci 46:10–15. https://doi.org/10.1016/j.cois.2021.01.005

Naik K, Mishra S, Srichandan H, Singh PK, Sarangi PK (2019) Plant growth promoting microbes: potential link to sustainable agriculture and environment. Biocatal Agric Biotechnol 21:101326. https://doi.org/10.1016/j.bcab.2019.101326

Neher DA, Barbercheck ME (2019) Soil microarthropods and soil health: intersection of decomposition and pest suppression in agroecosystems. Insects 10(12):414. https://doi.org/10.3390/insects10120414

Nishida H, Suzaki T (2018) Nitrate-mediated control of root nodule symbiosis. Curr Opin Plant Biol 44:129–136. https://doi.org/10.1016/j.pbi.2018.04.006

Norman JS, Smercina DN, Hileman JT, Tiemann LK, Friesen ML (2020) Soil aminopeptidase induction is unaffected by inorganic nitrogen availability. Soil Biol Biochem 149:107952. https://doi.org/10.1016/j.soilbio.2020.107952

Pan J, Huang C, Peng F, Zhang W, Luo J, Ma S, Xue X (2020) Effect of Arbuscular Mycorrhizal Fungi (AMF) and Plant Growth-Promoting Bacteria (PGPR) inoculations on *Elaeagnus angustifolia* L. in saline soil. Appl Sci 10(3):945. https://doi.org/10.3390/app10030945

Piron D, Boizard H, Heddadj D, Pérès G, Hallaire V, Cluzeau D (2017) Indicators of earthworm bioturbation to improve visual assessment of soil structure. Soil Tillage Res 173:53–63. https://doi.org/10.1016/j.still.2016.10.013

Porter SS, Sachs JL (2020) Agriculture and the disruption of plant–microbial symbiosis. Trends Ecol Evol 35:426–439. https://doi.org/10.1016/j.tree.2020.01.006

Potatov AM, Goncharov AA, Semenina EE, Korotkvich AY, Tsurikov SM, Rozanova OL, Anichkin AE, Zuey AG, Samoyolya SS, Semenyuk II, Yevdokimov IV, Tiunoy AV (2017) Arthropods in the subsoil: abundance and vertical distribution as related to soil organic matter, microbial biomass and plant roots. Eur J Soil Biol 82:88–97. https://doi.org/10.1016/j.ejsobi.2017.09.001

Potatov AM, Klarner B, Sandmann D, Widvastuti R, Scheu S (2019) Linking size spectrum, energy flux and trophic multifunctionality in soil food webs of tropical land-use systems. J Ecol Animals 88(12):1845–1859. https://doi.org/10.1111/1365-2656.13027

Ramos AC, Lima PT, Dias PN, Kasuya MCM, Feijó JA (2009) A pH signaling mechanism involved in the spatial distribution of calcium and anion fluxes in ectomycorrhizal roots. New Phytol 181:448–462

Recorbet G, Calabrese S, Balliuau T, Zivy M, Wipf D, Boller T, Courty PE (2021) Proteome adaptations under contrasting soil phosphate regimes of *Rhizophagus irregularis* engaged in a common mycorrhizal network. Fungal Genet Biol 147:103517. https://doi.org/10.1016/j. fgb.2021.103517

Ren C, Zhao F, Shi Z, Chen S, Han X, Yang G, Feng Y, Ren G (2017) Differential responses of soil microbial biomass and carbon-degrading enzyme activities to altered precipitation. Soil Biol Biochem 115:1–10. https://doi.org/10.1016/j.soilbio.2017.08.002

Rodriguez L, Suárez JC, Pulleman M, Guaca L, Rico A, Romero M, Quintero M, Lavelle P (2021) Agroforestry systems in the Colombian Amazon improve the provision of soil ecosystem services. Appl Soil Ecol 164:103933. https://doi.org/10.1016/j.apsoil.2021.103933

Rutten PJ, Poole PS (2019) Oxygen regulatory mechanisms of nitrogen fixation in rhizobia. In: Pooele RK (ed) Advances in microbial physiology. Elsevier. 325–389. https://doi.org/10.1016/bs.ampbs.2019.08.001

Rutten PJ, Steel H, Hood GA, Ramachadram VK, McMurtry L, Geddes B, Papachristodoulou AP, Poole PS (2021) Multiple sensors provide spatiotemporal oxygen regulation of gene expression in a *Rhizobium*-legume symbiosis. PLoS Genet. https://doi.org/10.1371/journal.pgen.1009099

Sagi N, Hawlena D (2021) Arthropods as the engine of nutrient cycling in arid ecosystems. Insects 12(8):726. https://doi.org/10.3390/insects12080726

Sagi N, Zaguri M, Hawlena D (2021) Macro-detritivores assist resolving the dryland decomposition conundrum by engineering an underworld heaven for decomposers. Ecosystems 24:56–67. https://doi.org/10.1007/s10021-020-00504-9

Sakamoto K, Ogiwara N, Kaji T, Sugimoto Y, Ueno M, Sonoda M, Matsui A, Ishida J, Tanaka M, Totoki Y, Shinozaki K, Saki M (2019) Transcriptome analysis of soybean (*Glycine max*) root genes differentially expressed in rhizobial, arbuscular mycorrhizal, and dual symbiosis. J Plant Res 132:541–568. https://doi.org/10.1007/s10265-019-01117-7

Sankaranarayanan C, Singaravelu B, Rajeshkumar M (2019) Entomopathogenic nematodes (EPN): diversity in Indian tropical sugarcane ecosystem and its biocontrol potential against white grub *Holotrichia serrata* F. on sugarcane. Sugar Tech 21:371–382. https://doi.org/10.1007/s12355-018-0628-9

Santos ADA, Silveira JAGD, Guilherme EDA, Bonifacio A, Rodrigues AC, Figueiredo MDVB (2018) Changes induced by co-inoculation in nitrogen–carbon metabolism in cowpea under salinity stress. Braz J Microbiol 49:685–694. https://doi.org/10.1016/j.bjm.2018.01.007

Schwember AR, Schulze J, del Pozo A, Cabeza RA (2019) Regulation of symbiotic nitrogen fixation in legume root nodules. Plan Theory 8(9):333. https://doi.org/10.3390/plants8090333

Shi K, Su T, Wang Z (2019) Comparison of poly (butylene succinate) biodegradation by *Fusarium solani* cutinase and *Candida antarctica* lipase. Polym Degrad Stab 164:55–60. https://doi.org/10.1016/j.polymdegradstab.2019.04.005

Silva RJ, Pelissari TD, Krinski D, Canale G, Vaz-de-Melo FZ (2017) Abrupt species loss of the Amazonian dung beetle in pastures adjacent to species-rich forests. J Insect Conserv 21:487–494. https://doi.org/10.1007/s10841-017-9988-9

Sofo A, Elshafie HS, Camele I (2020a) Structural and functional organization of the root system: a comparative study on five plant species. Plan Theory 9(10):1338. https://doi.org/10.3390/plants9101338

Sofo A, Mininni AN, Ricciuti P (2020b) Soil macrofauna: a key factor for increasing soil fertility and promoting sustainable soil use in fruit orchard agrosystems. Agronomy 10(4):456. https://doi.org/10.3390/agronomy10040456

Sorde KL, Ananthanarayan L (2019) Isolation, screening, and optimization of bacterial strains for novel transglutaminase production. Prep Biochem Biotechnol 49:64–73. https://doi.org/10.108 0/10826068.2018.1536986

Souza TAF, Santos D (2018) Biologia dos Solos da Caatinga. Universidade Federal da Paraíba, PPGCS, Areia

Souza TAF (2015) Handbook of arbuscular mycorrhizal fungi. Springer, Cham. https://doi.org/10.1007/978-3-319-24850-9

Sprunger CD, Culman SW, Peralta AL, DuPont T, Lennon JT, Snapp SS (2019) Perennial grain crop roots and nitrogen management shape soil food webs and soil carbon dynamics. Soil Biol Biochem 137:107573. https://doi.org/10.1016/j.soilbio.2019.107573

Suleiman AKA, Harkes P, van der Elsen S, Holterman M, Korthals GW, Helder J, Kuramae EE (2019) Organic amendment strengthens interkingdom associations in the soil and rhizosphere of barley (*Hordeum vulgare*). Sci Total Environ 695:133885. https://doi.org/10.1016/j.scitotenv.2019.133885

Thakur MP, Geisen S (2019) Trophic regulations of the soil microbiome. Trends Microbiol 27(9):771–780. https://doi.org/10.1016/j.tim.2019.04.008

Tuma J, Eggleton P, Fayle TM (2019a) Ant-termite interactions: an important but under-explored ecological linkage. Biol Rev 95(3):555–572. https://doi.org/10.1111/brv.12577

Tuma J, Fleiss S, Eggleton P, Frouz J, Klimes P, Lewis OT, Yusah KM, Fayle TM (2019b) Logging of rainforest and conversion to oil palm reduces bioturbator diversity but not levels of bioturbation. Appl Soil Ecol 144:123–133. https://doi.org/10.1016/j.apsoil.2019.07.002

Unuofin JO, Okoh AI, Nwodo UU (2019) Aptitude of oxidative enzymes for treatment of wastewater pollutants: a laccase perspective. Molecules 24(11):2064. https://doi.org/10.3390/molecules24112064

Uroz S, Coutry PE, Orger P (2019) Plant symbionts are engineers of the plant-associated microbiome. Trends Plant Sci 24(10):905–916. https://doi.org/10.1016/j.tplants.2019.06.008

Urrea-Galeano LA, Rosamond EA, Coates R, Mora F, Guillermo A, Marríquez I (2019) Dung beetle activity affects rain forest seed bank dynamics and seedling establishment. Biotropica 51(2):186–195. https://doi.org/10.1111/btp.12631

Vanolli BS, Canisares LP, Franco ALC, Delabie JHC, Cerri CEP, Cherubin MR (2021) Epigeic fauna (with emphasis on ant community) response to land-use change for sugarcane expansion in Brazil. Acta Oecol 110:103702. https://doi.org/10.1016/j.actao.2021.103702

Vara S, Karnena MK (2020) Fungal enzymatic degradation of industrial effluents – a review. Curr Res Environ Appl Mycol 10(1):417–442. https://doi.org/10.5943/cream/10/1/33

Veldhuis MP, Berg MP, Loreau M, Olff H (2018) Ecological autocatalysis: a central principle in ecosystem organization? Ecol Monogr 88(3):304–319. https://doi.org/10.1002/ecm.1292

Verma R, Annapragada H, Katiyar N, Shrutika N, Das K, Murugesan S (2020) Rhizobium. In: Amaresan N, Kumar SM, Annapurna K, Kumar K, Sankaranarayanan A (eds) Beneficial microbes in agro-ecology, 37–54. https://doi.org/10.1016/B978-0-12-823414-3.00004-6

Wan N, Ji X, Deng J, Kiær Cai Y, Jiang J (2019) Plant diversification promotes biocontrol services in peach orchards by shaping the ecological niches of insect herbivores and their natural enemies. Ecol Indic 99:387–392. https://doi.org/10.1016/j.ecolind.2017.11.047

Wan NF, Zheng XR, Fu LW, Kiær LP, Zhang Z, Chapli-Kramer R, Dainese M, Tan J, Qiu S, Hu Y, Tian W, Nie M, Ju R, Deng J, Jiang J, Cai Y, Li B (2020a) Global synthesis of effects of plant species diversity on trophic groups and interactions. Nat Plants 6:503–510. https://doi.org/10.1038/s41477-020-0654-y

Wan W, Wang Y, Tan J, Qin Y, Zuo W, Wu H, He H, He D (2020b) Alkaline phosphatase-harboring bacterial community and multiple enzyme activity contribute to phosphorus transformation during vegetable waste and chicken manure composting. Bioresour Technol 297:122406. https://doi.org/10.1016/j.biortech.2019.122406

Wang X, Feng H, Wang Y, Wang M, Xie X, Chang H, Wang L, Qu J, Sun K, He W, Wang C, Dai C, Tian C, Yu N, Zhang X, Liu H, Wang E (2021) Mycorrhizal symbiosis modulates the rhizosphere microbiota to promote rhizobia–legume symbiosis. Mol Plant 14(3):503–516. https://doi.org/10.1016/j.molp.2020.12.002

Wiesmeier M, Urbanski L, Hobley E, Lang B, von Lützow M, Marin-Spiotta E, Wesemael B, Rabot E, Liess M, Garcia-Franco N, Wollschäger U, Vogel H, Kogel-Knabner I (2019) Soil organic carbon storage as a key function of soils – a review of drivers and indicators at various scales. Geoderma 333(149):162. https://doi.org/10.1016/j.geoderma.2018.07.026

Wilkerson A, Olapade OA (2020) Relationships between organic matter contents and bacterial hydrolytic enzyme activities in soils: comparisons between seasons. Curr Microbiol 77:3937–3944. https://doi.org/10.1007/s00284-020-02223-9

Winagraski E, Kaschuk G, Monteiro PHR, Auer CG, Higa AR (2019) Diversity of arbuscular mycorrhizal fungi in forest ecosystems of Brazil: a review. Cerne 25:25–35. https://doi.org/10.1590/01047760201925012592

Wutzler T, Zaehle S, Schrumpf M, Aharens B, Reichstein M (2017) Adaptation of microbial resource allocation affects modelled long term soil organic matter and nutrient cycling. Soil Biol Biochem 115:322–336. https://doi.org/10.1016/j.soilbio.2017.08.031

Yang G, Wagg C, Veresoglou SD, Hempel S, Rilling MC (2018) How soil biota drive ecosystem stability. Trends Plant Sci 23(12):1057–1067. https://doi.org/10.1016/j.tplants.2018.09.007

Yang Q, Siemann E, Harvey JA, Ding J, Biere A (2021) Effects of soil biota on growth, resistance and tolerance to herbivory in *Triadica sebifera* plants. Geoderma 402:115191. https://doi.org/10.1016/j.geoderma.2021.115191

Zanetti ME, Blanco F, Reynoso M, Crespi M (2020) To keep or not to keep: mRNA stability and translatability in root nodule symbiosis. Curr Opin Plant Biol 56:109–117. https://doi.org/10.1016/j.pbi.2020.04.012

Zhang X, Ferris H, Mitchell J, Liang W (2017) Ecosystem services of the soil food web after long-term application of agricultural management practices. Soil Biol Biochem 111:36–43. https://doi.org/10.1016/j.soilbio.2017.03.017

# Chapter 4
# Plant-Soil Feedback

**Abstract** A feedback is an event that occurs when the output of an organism metabolism (e.g., litter deposition by the natural process of plant senescence) is used as input back into the soil food web as an energy resource for other organism (e.g., litter transformers – Scarabaeidae and Spirobolida) as part of a chain of cause and effect. This chapter will introduce the main differences between feedback and interaction into soil ecology and soil biology. Primary producers may alter soil ecosystem through litter deposition, rootability, and rhizodeposition. In turns, soil organisms are influenced to act as herbivores (e.g., when feeding the fresh plant tissues), decomposers (e.g., when decomposing the dead plant tissues), and symbionts (e.g., when colonizing the roots through arbuscules and root nodules formation). Here, we will examine the various pathways that primary producers influence soil properties, the plant-arthropod interactions, and soil organisms influence primary producers. Plant-soil feedback over long timescales appears to be important drivers of soil sustainability or soil disturbance levels.

**Keywords** Primary producers as soil conditioners · Positive plant-soil feedback · Negative plant-soil feedback · Plant-arthropod interactions · Soil sustainability

**Questions Covered in the Chapter**
1. What does feedback mean in soil biology and soil ecology?
2. What is the difference between feedback and interaction?
3. How do primary producers influence soil biological properties?
4. How do herbivores and predators influence plant-arthropod interactions?
5. How do ecosystem engineers promote soil structure?
6. How do soil organisms influence plant communities?

© The Author(s), under exclusive license to Springer Nature Switzerland AG 2022     55
T. Souza, *Soil Biology in Tropical Ecosystems*, https://doi.org/10.1007/978-3-031-00949-5_4

## 4.1   Introduction

Soil ecologists define feedback as an event that occurs when the output of an organism metabolism (e.g., litter deposition by the natural process of plant senescence) is used as input back into the soil food web as an energy resource for other organism (e.g., litter transformers – Scarabaeidae and Spirobolida) and as part of a chain of cause and effect (Souza and Santos 2018). Thus, we can classify two situations of feedback into soil food web:

1. Positive: An increase in output energy reinforces the input energy. Process such as resilience and resistance on soil organism community may naturally occur. The original chain remains. It reflects sustainability.
2. Negative: A rise in output energy decreases the input energy. A new chain is created. The functional redundancy prevails. It reflects disturbance.

On the other hand, interaction is a kind of action that occurs as two or more organisms influence one another (Bennett and Klironomos 2019). Here, it is important to clarify that in soil biology and soil ecology the words feedback and interaction describe two dissimilar things (Freilich et al. 2018; Giron et al. 2018; Picot et al. 2019). As already described, there are two kinds of feedback, i.e., positive and negative, while interaction is closely related to interactivity and interconnectivity and can be classified according to the main subject area (Table 4.1). In a specific feedback, we can find more than two interaction types working actively into the

**Table 4.1**  List of the main interaction types that occur into soil ecosystem

| Subject | Interaction type |
|---|---|
| Physics | Electromagnetic interactions |
| | Gravitational interactions |
| Chemistry | Molecular recognition |
| | Intermolecular force |
| | Noncovalent bonding |
| | Hydrogen bond |
| | Van der Walls force |
| | Aromatic interaction |
| | Cation-pi interaction |
| | Hydrophobic effect |
| | Coordination complex |
| Biochemistry | Protein interactions |
| | Interactome |
| | Drug interaction |
| | Synergistic interaction |
| | Antagonistic interaction |
| Biology | Biological interaction |
| | Gene-environment interaction |
| | Cell-cell interaction |

Adapted from Jiang et al. (2017), Sharifi et al. (2017), Bausch-Fluck et al. (2019), Gessner et al. (2019), Kovács et al. (2019), Afkhami et al. (2020) Restuccia and Tello-Ortiz (2020), Ferretti et al. (2021), Wu et al. (2021)

original (e.g., cation exchange capacity, soil pH, hydrolysis, enzymatic activity, symbiosis, reproduction, and metabolism). However, describing a feedback as only one type of interaction is the biggest mistake that a soil ecologist can do during a study about functional ecology, conservation, or disturbance degree (Sun et al. 2018). We must understand that only one interaction will never create a plant-soil feedback by their own. A plant-soil feedback is a complex chain of cause and effect. It involves specific sequences of interactions. Into this chapter, we will illustrate the main pathways and the linkages between plant species and soil organisms into the tropical soils.

## 4.2  Plant Species Influencing Soil Biological Properties

Plant species can influence soil biological properties through litter deposition, root-ability, and rhizodeposition (Tian et al. 2019; Forstall-Sosa et al. 2020; Pandey et al. 2020). These processes can increase both food and habitat provision to many soil organisms (Hugoni et al. 2018; Briones et al. 2019). It is the main process related to a positive plant-soil feedback, and accordingly to Silva et al. (2018), disturbed areas commonly show a significant decrease in litter deposition (e.g., by low plant growth and biomass production), and it decreases in a chain of events reduction in root growth which directly reduces both rootability and rhizodeposition (D'Antonio et al. 2017). Litter deposition will affect directly macrofauna activity and soil organic matter contents. On the other hand, the rootability and rhizodeposition will affect root growth and soil C-rich compounds content, receptively. Finally, the sum of these three processes will affect microbiota activity. It promotes nutrient cycling and plant growth, thus creating a positive plant-soil feedback (Fig. 4.1).

We must consider that the litter deposition is the first and major source of plant C input to soil ecosystem (Martínez-Atencia et al. 2020). It occurs on soil surface and provides resource and conditions to soil organic horizon formation (Soko et al. 2018) as already described in Chap. 1. This process may influence soil biological properties through energy input and nutrient supply for litter transformers and decomposers. Also, it promotes habitat provision for many soil organisms. Rootability is the second source of plant C input to soil ecosystem (Qi et al. 2019). It occurs in the rhizosphere, and it defines the root growth and the ability to produce and release exudates that have the properties to interact with soil colloids (e.g., clay) and soil organisms (e.g., stimulating AMF presymbiotic phase) (Cheng et al. 2018; Laurindo et al. 2021). Finally, rhizodeposition is the third source of plant C input to soil ecosystem. It occurs around the rhizosphere, and it defines the sum of all C-rich compounds (e.g., water-soluble exudates, secretions of insoluble materials, lysates, dead fine roots, and gases) lost from plant roots (Bowsher et al. 2018; Hirte et al. 2018; Hassan et al. 2019).

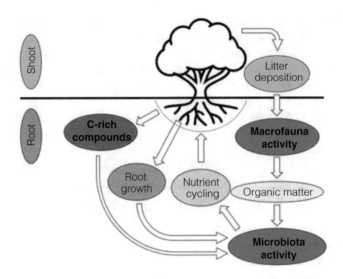

**Fig. 4.1** Schematic view of a positive plant-soil feedback. The primary producer can affect soil organisms and soil ecosystem through three main processes. (Adapted from Prescott and Vesterdal (2021), Wang et al. (2021))

## 4.3   Plant-Arthropod Interactions

We can find many complex relationships between plants and animals around all the world. However, the most important interaction is found between flowering plants and arthropods. As already described in the last chapter, the interaction between plant and soil organisms classified as herbivores (e.g., arthropods) is the most ancient ecological interaction in the humankind history (Currano et al. 2021). Some of these interactions are mutually beneficial (e.g., pollination or even the herbivory), while others are detrimental and in certain cases lethal to the plant species (Liu et al. 2019; Franco et al. 2020). The triggers which define if the interaction will be beneficial or detrimental are the food availability and the presence of predators as described in the niche hypothesis (in't Zandt et al. 2019).

To better understand the niche hypothesis, we must clearly describe both biotic and abiotic factors that affect plant-arthropod interaction (Losapio et al. 2020). In the biotic factors, we must include the food availability (e.g., defined by the fresh plant tissue – leaves) and abundance of predators (e.g., Araneidae, Filistatidae, Forficulidae, and Vespidae), while in the abiotic factors we must include thermal amplitude, soil nutrient availability (e.g., high soil N content tends to attract plant pathogens), and other non-living factors (Prather et al. 2020). These factors are relevant to define the kind of interaction between plant species and herbivores. If the soil ecosystem provides enough conditions to sustain a high diversity of primary producers, it will sustain a well-defined food web with many predators promoting biological control. However, in agricultural soils where the monodominance prevails, the low diversity of primary producers promotes an environment where the birth rate of predators is significantly reduced, and the herbivores food requirement increases drastically (See the consumer-resource model) thus creating a negative plant-soil feedback (Emery et al. 2021).

## 4.4 Soil Structure

Attempts have been made to provide many frameworks to explain how soil biota affect soil structure, based on plant-soil feedback (Rinella and Reinhart 2018). It is well-established among soil ecologist that soil organisms adapted to promote soil structure are classified as ecosystem engineers, and they have different sets of traits to provide ecosystem services (Lehmann et al. 2017; Arnol et al. 2019; McCary and Schmitz 2021), and it is becoming clear that these sets of traits can also determine soil structure, especially in organic and A horizons. Formicidae is a group of eusocial insects that acts as superorganisms making galleries and nests into soil profile. They promote soil porosity and soil aggregation constantly through their behavior. Also, they provide favorable conditions to root growth, Haplotaxida, and other vermes mobilization (Elizalde et al. 2020). Another example is the dung beetle (Scarabaeidae) that consumes dung both in larval and final forms. They store dung balls in their nests full of eggs (Tonelli 2021). This improves food availability to their larvae and improves soil organic carbon content, soil aeration, and nutrient cycling by releasing essential nutrients to other soil organisms (Marriote et al. 2018; Maldonado et al. 2019; Kaleri et al. 2020).

## 4.5 Plant Communities Influenced by Soil Biota

Understanding the variables that cause shifts into the plant community composition in natural and disturbed ecosystems at the tropics has long been a major theme into soil biology and soil ecology (Valencia et al. 2018; Leff et al. 2018; Tuma et al. 2020; Walter et al. 2020). As already described, a wide variety of abiotic and biotic factors have been identified as being key factors of shifts in plant community composition. The plant-soil feedback is the result of the fitness of plant species, soil organisms, and their interactions. In fact, we can classify three main pathways where soil biota can strongly promote changes into plant community:

1. The activity of decomposer on the dead plant materials: It promotes changes into the nutrient cycling process. If we found a high quantity and quality of dead plant materials on soil surface, we can expect a positive plant-soil feedback and positive changes into plant community composition.

2. The activity of soil organisms on living plant roots: It promotes changes into the rhizosphere environment, thus improving plant nutrient and water uptake, herbivory process, and even decomposers.
3. The microbiota activity on soil organic matter: It improves nutrient cycling, thus promoting nutrient use by plants.

Another interesting process, where we can identify an entire plant community being affected by a soil organism, is the mycorrhizal association. This phenomenon benefits the performance of the plant species due to improving plant nutrient and water uptake and increasing plant resistance against root-feeding nematodes and bacteria (Dowarah et al. 2021; Mitra et al. 2021). In disturbed ecosystem under biological invasion process, the mycorrhizal association must promote the invasive plant species through different degrees of benefit on plant performance as described by Souza et al. (2019). The presence of specific arbuscular mycorrhizal fungi (e.g., *Funneliformis mosseae*, *Rhizoglomus intraradices*, and *Claroideoglomus claroideum*) led to a shift in plant community composition, reducing plant diversity in some cases (Souza et al. 2018; Chen et al. 2019). This suggests that arbuscular mycorrhizal fungi can reduce plant diversity in disturbed environments at the tropics. Similarly, the ectomycorrhizal relationship promotes the dominance of certain plant species at the tropics (Corrales et al. 2018).

## 4.6   Conclusions

There are evidence showing that the understanding of the plant-soil feedback is important to explain how the interactions between plant communities and soil ecosystem work. The plant-soil feedback can be defined as positive or negative, and we must consider that these two kinds of feedbacks are generated by more than two interaction between soil organisms (e.g., here including primary producers). The soil biological properties can be influenced by plant species through litter deposition (the main source of C on soil surface), rootability, and rhizodeposition. On the other hand, plant-arthropod interaction is modulated by two conditions: food availability and the abundance of predators. Finally, the role of the ecosystem engineers on soil structure. Over long timescales, plant-soil feedback appears to be important driver of plant community composition, and long-term changes in any of these factors/variables described into this chapter determine ecosystem productivity, soil organism activity, ecosystem services, and community composition at the tropics.

## References

Afkhami M, Almeida BK, Hernandez DJ, Kiesewetter KN, Revillini DP (2020) Tripartite mutualisms as models for understanding plant–microbial interactions. Curr Opin Plant Biol 56:28–36. https://doi.org/10.1016/j.pbi.2020.02.003

Arnol D, Schapiro D, Bodenmiller B, Saez-Rodriguez J, Stegle O (2019) Modeling cell-cell interactions from spatial molecular data with spatial variance component analysis. Cell Rep 29(1):202–211. https://doi.org/10.1016/j.celrep.2019.08.077

Bausch-Fluck D, Milani ES, Wollscheid B (2019) Surfaceome nanoscale organization and extracellular interaction networks. Curr Opin Chem Biol 48:26–33. https://doi.org/10.1016/j.cbpa.2018.09.020

Bennett JA, Klironomos J (2019) Mechanisms of plant–soil feedback: interactions among biotic and abiotic drivers. New Phytol 222(1):91–96. https://doi.org/10.1111/nph.15603

Bowsher AW, Evans S, Tiemann LK, Friesen ML (2018) Effects of soil nitrogen availability on rhizodeposition in plants: a review. Plant Soil 423:59–85. https://doi.org/10.1007/s11104-017-3497-1

Briones MJI, Elias DMO, Grant HK, McNamara NP (2019) Plant identity control on soil food web structure and C transfers under perennial bioenergy plantations. Soil Biol Biochem 138:107603. https://doi.org/10.1016/j.soilbio.2019.107603

Chen Q, Wu W, Qi S, Cheng H, Li Q, Ran Q, Dai ZC, Du DL, Egan S, Thomas T (2019) Arbuscular mycorrhizal fungi improve the growth and disease resistance of the invasive plant *Wedelia trilobata*. J Appl Microbiol 130(2):582–591. https://doi.org/10.1111/jam.14415

Cheng N, Peng Y, Kong Y, Li J, Sun C (2018) Combined effects of biochar addition and nitrogen fertilizer reduction on the rhizosphere metabolomics of maize (*Zea mays* L.) seedlings. Plant Soil 433:19–35. https://doi.org/10.1007/s11104-018-3811-6

Corrales A, Henkel TW, Smith ME (2018) Ectomycorrhizal associations in the tropics – biogeography, diversity patterns and ecosystem roles. New Phytol 220(4):1076–1091. https://doi.org/10.1111/nph.15151

Currano ED, Azevedo-Schmidt LE, Maccraken SA, Swain A (2021) Scars on fossil leaves: an exploration of ecological patterns in plant–insect herbivore associations during the age of angiosperms. Palaeogeogr Palaeoclimatol Palaeoecol 582:110636. https://doi.org/10.1016/j.palaeo.2021.110636

D'Antonio CM, Yelenik SG, Mack MC (2017) Ecosystem vs. community recovery 25 years after grass invasions and fire in a subtropical woodland. J Ecol 105(6):1462–1474. https://doi.org/10.1111/1365-2745.12855

da Silva BW, Périco E, Dalzochio MS, Santos M, Cajaiba RL (2018) Are litterfall and litter decomposition processes indicators of forest regeneration in the neotropics? Insights from a case study in the Brazilian Amazon. For Ecol Manag 429:189–197. https://doi.org/10.1016/j.foreco.2018.07.020

Dowarah B, Gill SS, Agarwala N (2021) Arbuscular mycorrhizal fungi in conferring tolerance to biotic stresses in plants. J Plant Growth Regulat. https://doi.org/10.1007/s00344-021-10392-5

Elizalde L, Arbertman M, Arnan X, Eggleton P, Leal IR, Lescano MN, Saez A, Werenkraut V, Pirk GI (2020) The ecosystem services provided by social insects: traits, management tools and knowledge gaps. Biol Rev 95(5):1418–1441. https://doi.org/10.1111/brv.12616

Emery SE, Jonsson M, Silva H, Ribeiro A, Mills NJ (2021) High agricultural intensity at the landscape scale benefits pests, but low intensity practices at the local scale can mitigate these effects. Agric Ecosyst Environ 306:107199. https://doi.org/10.1016/j.agee.2020.107199

Ferretti A, Prampolini G, Ischia M (2021) Noncovalent interactions in catechol/ammonium-rich adhesive motifs: reassessing the role of cation-$\pi$ complexes? Chem Phys Lett 779:138815. https://doi.org/10.1016/j.cplett.2021.138815

Forstall-Sosa KS, Souza TAF, Lucena EO, Silva SAI, Ferreira JTA, Silva TN, Santos D, Niemeyer JC (2020) Soil macroarthropod community and soil biological quality index in a green manure farming system of the Brazilian semi-arid. Biologia. https://doi.org/10.2478/s11756-020-00602-y

Franco ALC, Gherardi LA, de Tomasel CM, Andriuzzi WS, Ankrom KE, Bach EM, Guan P, Sal OE, Wall DH (2020) Root herbivory controls the effects of water availability on the partitioning between above and belowground grass biomass. Funct Ecol 34(11):2403–2410. https://doi.org/10.1111/1365-2435.13661

Freilich MA, Wieters E, Broitman BR, Marquet PA, Navarrete SA (2018) Species co-occurrence networks: can they reveal trophic and non-trophic interactions in ecological communities? Ecology 99(3):690–699. https://doi.org/10.1002/ecy.2142

Gessner A, König J, Fromm MF (2019) Clinical aspects of transporter-mediated drug–drug interactions. Clin Pharmacol Therap 105(6):1386–1394. https://doi.org/10.1002/cpt.1360

Giron D, Dubreuil G, Bennett A, Dedeine F, Dicke M, Dyer LA, Erb M, Harris MO, Huguet E, Kaloshian I, Kawakita A, Lopez-Vaamonde C, Palmer T, Petanidou T, Poulsen M, Salle A, Simon J, Terblanche JS, Thiery D, Whiteman NK, Woods HA, Pincebourde S (2018) Promises and challenges in insect–plant interactions. Entomol Exp Appl 166(5):319–343. https://doi.org/10.1111/eea.12679

Hassan MK, McInroy JA, Kloepper JW (2019) The interactions of rhizodeposits with plant growth-promoting rhizobacteria in the rhizosphere: a review. Agriculture 9(7):142. https://doi.org/10.3390/agriculture9070142

Hirte J, Leifeld J, Abiven S, Oberholzer H, Mayer J (2018) Below ground carbon inputs to soil via root biomass and rhizodeposition of field-grown maize and wheat at harvest are independent of net primary productivity. Agric Ecosyst Environ 265:556–566. https://doi.org/10.1016/j.agee.2018.07.010

Hugoni M, Luis P, Guyonnet J, Haichar F et al (2018) Plant host habitat and root exudates shape fungal diversity. Mycorrhiza 28:451–463. https://doi.org/10.1007/s00572-018-0857-5

In't Zandt D, van den Brink A, de Kroon H, Visser EJW (2019) Plant-soil feedback is shut down when nutrients come to town. Plant Soil 439:541–551. https://doi.org/10.1007/s11104-019-04050-9

Jiang L, Cao S, Cheung PPH, Zheng X, Leung CWT, Peng Q, Shuai Z, Tang BZ, Yao S, Huang X (2017) Real-time monitoring of hydrophobic aggregation reveals a critical role of cooperativity in hydrophobic effect. Nat Commun 8:1–8. https://doi.org/10.1038/ncomms15639

Kaleri AR, Ma J, Jakhar AM, Hakeem A, Ahmed A, Napar WPF, Ahmed S, Han Y, Abro SA, Nabi F, Tan C, Kaleri AH (2020) Effects of dung beetle-amended soil on growth, physiology, and metabolite contents of Bok choy and improvement in soil conditions. J Soil Sci Plant Nutr 20(4):2671–2683. https://doi.org/10.1007/s42729-020-00333-8

Kovács IA, Luck K, Spirohn K, Wang Y, Pollis C, Schalabach S, Bian W, Kim D, Kishore N, Hao T, Calderwood MA, Vidal M, BArabási A (2019) Network-based prediction of protein interactions. Nat Commun 10:1240. https://doi.org/10.1038/s41467-019-09177-y

Laurindo LK, Souza TAF, da Silva LJR, Casal TB, Pires KJC, Kormann S, Schmitt DE, Siminski A (2021) Arbuscular mycorrhizal fungal community assembly in agroforestry systems from the Southern Brazil. Biologia 76(4):1099–1107. https://doi.org/10.1007/s11756-021-00700-5

Leff JW, Bardgett RD, Wilkinson A, Jackson BG, Pritchard WL, de Long JR, Oakley S, Mason KE, Ostle NJ, Johnson D, Baggs EM, Fierer N (2018) Predicting the structure of soil communities from plant community taxonomy, phylogeny, and traits. ISME J 12:1794–1805. https://doi.org/10.1038/s41396-018-0089-x

Lehmann A, Zheng W, Rillig MC (2017) Soil biota contributions to soil aggregation. Nat Ecol Evol 1(12):1828–1835. https://doi.org/10.1038/s41559-017-0344-y

Liu H, Macdonald CA, Cook J, Anderson IC, Singh BK (2019) An ecological loop: host microbiomes across multitrophic interactions. Trends Ecol Evol 34(12):1118–1130. https://doi.org/10.1016/j.tree.2019.07.011

Losapio G, Shcmid B, Bascompte J, Pierfilippo RM, Christoph C, Haenni GJ, Neumeyer R (2020) An experimental approach to assessing the impact of ecosystem engineers on biodiversity and ecosystem functions. Ecology 102(2):e03243. https://doi.org/10.1002/ecy.3243

Maldonado MB, Aranibar JN, Serrano AM, Chacoff NP, Vázquez DP (2019) Dung beetles and nutrient cycling in a dryland environment. Catena 179:66–73. https://doi.org/10.1016/j.catena.2019.03.035

Marriote P, Mehrabi Z, Bezemer TM, de DeynGB KA, Drigo B, Veen CGF, van der Heijden MGA, Kardol P (2018) Plant–soil feedback: bridging natural and agricultural sciences. Trends Ecol Evol 33(2):129–142. https://doi.org/10.1016/j.tree.2017.11.005

Martínez-Atencia J, Loaiza-Usuga JC, Osorio-Veja NW, Correa-Londoño G, Casamitjana-Causa M (2020) Leaf litter decomposition in diverse silvopastoral systems in a neotropical environment. J Sustain For 39(7):710–729. https://doi.org/10.1080/10549811.2020.1723112

McCary MA, Schmitz OJ (2021) Invertebrate functional traits and terrestrial nutrient cycling: insights from a global meta-analysis. J Anim Ecol 90(7):1714–1726. https://doi.org/10.1111/1365-2656.13489

Mitra D, Djebaili R, Pellegrini M, Mahakur B, Sarker A, Chaudhary P, Khoshru B, Del Gallo M, Kitouni M, Barik DP, Panneerselvam P, das Mohapatra PK (2021) Arbuscular mycorrhizal symbiosis: plant growth improvement and induction of resistance under stressful conditions. J Plant Nutr 44(13):1993–2028. https://doi.org/10.1080/01904167.2021.1881552

Pandey VC, Rai A, Singh L, Singh DP (2020) Understanding the role of litter decomposition in restoration of fly ash ecosystem. Bull Environ Contaminat Toxicol. https://doi.org/10.1007/s00128-020-02994-8

Picot A, Monnin T, Loeuille MN (2019) From apparent competition to facilitation: impacts of consumer niche construction on the coexistence and stability of consumer-resource communities. Funct Ecol 33(9):1746–1757. https://doi.org/10.1111/1365-2435.13378

Prather RM, Castiollioni K, Welti EAR, Kaspari M, Souza L (2020) Abiotic factors and plant biomass, not plant diversity, strongly shape grassland arthropods under drought conditions. Ecology 101(6):e03033. https://doi.org/10.1002/ecy.3033

Prescott CE, Vesterdal L (2021) Decomposition and transformations along the continuum from litter to soil organic matter in forest soils. For Ecol Manag 498:119522. https://doi.org/10.1016/j.foreco.2021.119522

Qi Y, Wei W, Chen C, Chen L (2019) Plant root-shoot biomass allocation over diverse biomes: a global synthesis. Glob Ecol Conserv 18:e00606. https://doi.org/10.1016/j.gecco.2019.e00606

Restuccia A, Tello-Ortiz F (2020) Pure electromagnetic-gravitational interaction in Hořava–Lifshitz theory at the kinetic conformal point. Eur Phys J C 80:86. https://doi.org/10.1140/epjc/s10052-020-7674-7

Rinella MJ, Reinhart RO (2018) Toward more robust plant-soil feedback research. Ecology 99(3):550–556. https://doi.org/10.1002/ecy.2146

Sharifi P, Aminpanah H, Erfani R, Mohaddesi A, Abbasian A (2017) Evaluation of genotype× environment interaction in rice based on AMMI model in Iran. Rice Sci 24(3):173–180. https://doi.org/10.1016/j.rsci.2017.02.001

Soko NW, Mark JS, Bradford MA (2018) Pathways of mineral-associated soil organic matter formation: integrating the role of plant carbon source, chemistry, and point of entry. Glob Chang Biol 25:12–24. https://doi.org/10.1111/gcb.14482

Souza TAF, de Andrade LA, Freitas H, Sandim AS (2018) Biological invasion influences the outcome of plant-soil feedback in the invasive plant species from the Brazilian semi-arid. Microb Ecol 76:102–112. https://doi.org/10.1007/s00248-017-0999-6

Souza TAF, Santos (2018) Biologia dos Solos da Caatinga. Universidade Federal da Paraíba, PPGCS, Areia

Souza TAF, Santos D, de Andrade LA, Freitas H (2019) Plant-soil feedback of two legume species in semi-arid Brazil. Braz J Microbiol 50:1011–1020. https://doi.org/10.1007/s42770-019-00125-y

Sun G, Wang C, Chang L, Wu Y, Li L, Jin Z (2018) Effects of feedback regulation on vegetation patterns in semi-arid environments. Appl Math Model 61:200–215. https://doi.org/10.1016/j.apm.2018.04.010

Tian K, Kong X, Yuan L, Lin H, He Z, Yao B, Ji J, Sun S, Tian X (2019) Priming effect of litter mineralization: the role of root exudate depends on its interactions with litter quality and soil condition. Plant Soil 440:457–471. https://doi.org/10.1007/s11104-019-04070-5

Tonelli M (2021) Some considerations on the terminology applied to dung beetle functional groups. Ecol Entomol 46(4):772–776. https://doi.org/10.1111/een.13017

Tuma J, Eggleton P, Fayle TM (2020) Ant-termite interactions: an important but under-explored ecological linkage. Biol Rev 95(3):555–572. https://doi.org/10.1111/brv.12577

Valencia E, Gross N, Quero JL, Carmona CP, Ochoa V, Gozalo B, Delgado-Baquerizo M, Dumack K, Hamonts K, Singh BK, Bonkowski M, Maestre FT (2018) Cascading effects from plants to soil microorganisms explain how plant species richness and simulated climate change affect soil multifunctionality. Glob Chang Biol 24(12):5642–5654. https://doi.org/10.1111/gcb.14440

Walter J, Buchmann CM, Schurr FM (2020) Shifts in plant functional community composition under hydrological stress strongly decelerate litter decomposition. Ecol Evol 10(12):5712–5724. https://doi.org/10.1002/ece3.6310

Wang W, Hu K, Huang K, Tao J (2021) Mechanical fragmentation of leaf litter by fine root growth contributes greatly to the early decomposition of leaf litter. Glob Ecol Conserv 26:e01456. https://doi.org/10.1016/j.gecco.2021.e01456

Wu J, Hou Z, Lang J, Cui J, Yang J, Dai J, Li C, Yan Y, Xie A (2021) Facile preparation of metal-polyphenol coordination complex coated PVDF membrane for oil/water emulsion separation. Sep Purif Technol 258(2):118022. https://doi.org/10.1016/j.seppur.2020.118022

# Chapter 5
# Trophic Structure and Soil Biological Communities

**Abstract** This chapter explores current knowledge used to study the trophic structure and the soil biological communities at the tropics. Trophic structure is defined by the complex of various feeding levels (e.g., producers, decomposers, herbivores, etc.) into an entire community, while the soil biological communities represent the wide range of soil organisms (e.g., roots, insects, nematodes, bacteria, fungi, etc.) that live in soil profile. These two concepts are strongly linked to each other. Trophic groups can influence decomposition, aboveground herbivory, net primary production, and nutrient cycling. It has been recognized that roots are an important driver of soil biological communities and most recently have soil ecologists started describing the effects of root exudates on soil biota diversity and community composition. This chapter examines some of the many ways that root exudates, plant biomass production, and litter may influence trophic structure and soil biological communities, and these changes may in turn feedback to affect the soil ecosystem at the tropics.

**Keywords** Trophic groups · Autotrophs · Heterotrophs · Root exudates · Net primary production

**Questions Covered in the Chapter**
1. How do root exudates promote soil biota community?
2. Why is it important to consider the net primary production in soil biology?
3. What is the relationship among litter and trophic structure?

## 5.1 Introduction

Tropical ecosystems have their abiotic and biotic compartments very well characterized isolated to each other (Mishra et al. 2020; Stevens et al. 2020). However, the sizes and relationships between these compartments for store energy are not well-described (Klimešová et al. 2018; Nielsen et al. 2020; Glatthorn et al. 2021).

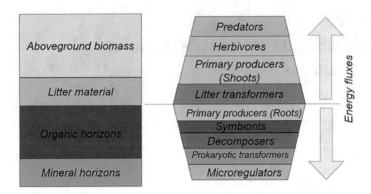

**Fig. 5.1** Schematic view of the soil ecosystem trophic structure. (Adapted from Souza and Freitas 2018)

The main characteristics shared between ecosystems are their trophic structure (Potatov et al. 2018; Jackson et al. 2019; Tsurikov et al. 2019). Trophic structure represents the wide variety of feeding levels into an entire community (Fig. 5.1). Basically, we can classify two main groups in a trophic structure:

1. Autotrophs: Where we can find all the organisms that synthetize their organic compounds from abiotic materials (e.g., $CO_2$ and $N_2$). They are capable of fixing abiotic carbon through photosynthesis in organic compounds such as carbohydrates. Thus, this group is very dependent to the solar energy to power the chemical synthesis of the abiotic carbon. Other examples of autotrophs are the free-living (e.g., *Azotobacter* and *Clostridium*) and symbiotic N-fixing bacteria (e.g., *Rhizobium*) that convert inorganic N into ammonia by ammonification. These bacteria are commonly descried as Diazotrophs. The soil organisms included into this group are the primary producers (plant species), autotrophic bacteria, and algae (Zhengtao et al. 2019; Mahmud et al. 2020; Liu et al. 2021; Morgan et al. 2021).

2. Heterotrophs: Where we find the primary, secondary, and tertiary consumers that obtain their energy and resources by feeding on autotrophs. They are unable to produce its own energy; thus we will never find a producer classified into this group. Detritivores and Saprotrophs are the most dominant organisms classified as heterotrophs. The first one obtains energy by consuming plant and animal parts, while the second one obtains energy by the decayed organic matter through extracellular digestion. The soil organisms included into this group are the Myriapods, Arachnids, Nematodes, Fungi, and Insects (Certini et al. 2021; Potatov et al. 2021).

On the other hand, the soil biological communities as already described in Chap. 2 are represented by a wide variety of soil organisms that differ in morphology, physiology, metabolism, and behavior. This kind of concept helps us to understand all the soil organism community. However, it does not show the soil organism function. We know that soil organism can do the following:

1. Growth by transforming inorganic compounds into extraordinarily complex organic structures, thus improving net primary production and soil heterotrophs community composition.
2. Transform and incorporate the litter and organic compounds in soil profile, thus improving soil structure and nutrient cycling and creating a positive plant-soil feedback.

Overall, soil organisms that can transform inorganic compounds into organic structures are classified as autotrophs (Tang et al. 2020; Tang et al. 2021). They can modify their environment by changing soil properties and the surrounding heterotroph community through root exudates (Tsunoda and van Dam 2017; Olanrewaju et al. 2019; Zhang et al. 2021a). These changes may create a positive plant-soil feedback thus promoting plant biomass production (Semchenko et al. 2017; Zhang et al. 2021b; Zhao et al. 2021), and considering a temporal scale, healthy plants in a tropical ecosystem (e.g., tropical rainforest) have a great impact on litter deposition (Marian et al. 2018; Vives-Peris et al. 2020). This chapter will highlight the following points:

1. The different mechanisms by which root exudates affect trophic structure and soil properties.
2. How plant biomass production affects the trophic structure at the tropical ecosystems.
3. How the litter operates into the trophic structure.

## 5.2   Root Exudates

There are two pathways by which root exudates can influence trophic structure and soil properties in soil profile (Eisenhauer et al. 2017; Rilfe et al. 2019; Sun et al. 2021). In the short term, root exudates can only influence trophic structure through root exudation patterns (e.g., $H^+$ extrusion, allelopathy, and C-rich compounds release), whereas in the long term, the root exudates can influence both physical and chemical properties by shifts in the trophic structure (e.g., positive plant-soil feedback) (Oburger and Jones 2018; Herz et al. 2019). Both these pathways are intimately linked and influence primary production and plant community composition (Zhang et al. 2017; Xia et al. 2019; Preece and Peñuelas 2020). Below, we described the $H^+$ extrusion, the allelopathy, and the rhizodeposition separately.

Plant species extrude the excess of $H^+$ generated by their metabolism on the rhizosphere zone. This zone is directly influenced by root exudation and indirectly by the associated soil microbiota community. The rhizosphere comprises the root microbiome (Fig. 5.2). In this microbiome, the $H^+$ extrusion process usually occurs in the roots through the nutrient uptake process (Marastoni et al. 2020; Siao et al. 2020). Overall, we must consider that the soil pH from a non-rhizospheric zone is always lower than the soil pH from a rhizospheric zone because of the intense release of $H^+$ by the roots. This phenomenon is higher in roots colonized by N-fixing

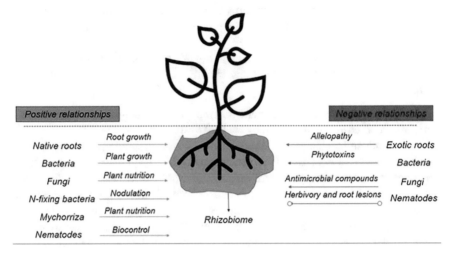

**Fig. 5.2** Root microbiome and their associated soil organism communities. (Adapted from Sharma and Verma (2018))

bacteria or even arbuscular mycorrhizal fungi (Igiehon and Babalola 2020; Cantó et al. 2020; da Silva et al. 2021), and it promotes rhizosphere acidification (e.g., by changing soil pH) thus creating optimal conditions to some soil organisms and to increase the availability of P, Mo, and Fe which are especially needed for both plant species and soil organism nutrition (Zhang et al. 2019; Robles-Aguilar et al. 2020; Mahmud et al. 2021). This shift in the soil pH around rhizosphere improves diazotrophs, nematodes, and mesofauna diversity. Once improving nutrient availability, it also contributes with the energy supply for a wide diversity of trophic levels and improves significantly with the net primary production.

The roots produce secretions (Table 5.1) classified as allelochemicals (e.g., also called secondary metabolites) that control seed germination, plant growth, and the survival rate of other organisms (Latif et al. 2017; Mehmood et al. 2019; Hickman et al. 2021).

These allelochemical compounds are involved in the allelopathy process, which can be positive if the biochemicals have beneficial effects (e.g., some allelochemicals promote the symbiosis among plant species and arbuscular mycorrhizal fungi) or be negative if the biochemicals have detrimental effects on other organisms' fitness (e.g., some allelochemicals have an antibiotic, anti-mutualism, and anti-herbivory effects) (Fig. 5.3). We must consider that these biochemicals are affected by both abiotic (e.g., soil pH), and biotic (e.g., P availability) factors, and they are an important driver that determines the abundance and distribution of soil organisms within plant communities, and it goes beyond the rhizosphere (Legrand et al. 2018; Barbato et al. 2019; Bluhm et al. 2019).

Thus, allelopathy defines all chemical interactions among organisms into a natural or an anthropized environment (Aslam et al. 2017; da Silva et al. 2017; Hussain

**Table 5.1** Typical root secretions produced in the root microbiome

| Group | Substance |
|---|---|
| Enzymes | Amylase, invertase, phosphatase, protease, polygalacturonase |
| Growth factors | p-Amino benzoic acid, auxins, biotin, choline, inositol, n-methyl nicotinic acid, niacin, pantothenate, pyridoxine, thiamine |
| Macromolecules | Nucleotide, flavanone, fatty acids, proteins, sterols, lipids, aliphatics, aromatics, carbohydrates |
| N-rich compounds | α- and β-Alanine, arginine, asparagine, aspartic acid, cystine, Cysteine, glutamine, glycine, histidine, lysine, methionine, phenylalanine, proline, serine |
| Organic acids | Acetic, citric, fumaric, lactic, malic, malonic, oxalic, succinic, tartaric |
| Phenolic acid | Caffeic acid, cinnamic acid, coumarin, ferulic acid, salycilic acid, syringic acid, vanillic acid |
| Sugar | Arabinose, fructose, fucose, galactose, glucose, maltose, oligosaccharide, raffinose, rhamnose, ribose, sucrose, xylose |

Adapted from Novo et al. (2018), Zhao et al. (2018)

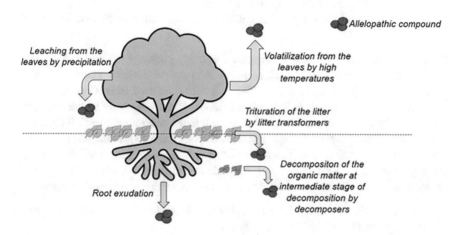

**Fig. 5.3** Main pathways of allelochemicals in soil profile. (Adapted from Souza et al. 2022)

et al. 2021). As already described, these interactions can be positive or negative, but all of them must present direct effect of an organism on another one (Arroyo et al. 2018; Scavo et al. 2019; Zhang et al. 2021a, b). Soil organisms such as plants, algae, bacteria, and fungi constantly influence the net primary production of soil ecosystem (Capek et al. 2018; Guo et al. 2019; Trivedi et al. 2020). Some examples of allelopathic effects are as follows:

1. Plant defense against herbivory:

    (a) Volatile terpenes produced by some shrubs that exclude rodents and other herbivores. It allows the grass to grow in the bare zones.

2. Plant against plant:

   (a) Seed bank germination suppression, which is quite common in biological invasion. It is an important process that determines the success of many invasive species such as *Prosopis juliflora*, *Cryptostegia madagascariensis*, and *Sesbania virgata* in many areas in the Brazilian Northeast.
   (b) Plant species inhibits the growth of neighboring plants (e.g., *Ailanthus altissima* and *Amaranthus sp.*).

3. Soil sickness:

   (a) In tropical fruit trees, we can find different types of deleterious microorganisms and nematodes involved in areas where plant species were grown repeatedly in the same field.

4. Preventing mutualisms:

   (a) Some glucosinolates (e.g., Sinigrin) produced by plant species from Brassicaceae can interfere within the symbiotic relationship between plant roots and arbuscular mycorrhizal fungi.
   (b) Some *Eucalyptus* species produce antibiotic substances that control certain soil microbes.

Finally, the rhizodeposition that occurs constantly during plant life cycle produces and releases various C- and N-rich compounds in the rhizosphere (Hupe et al. 2018; Henneron et al. 2020; Jeewani et al. 2020). It is also referred as root cell waste. The rhizodeposition provides energy supply for the soil biota community around the rhizosphere, and it affects the soil chemical properties in the rhizospheric zone. It allows the growth of soil organisms surrounding and inside plant roots. In general, this process can lead to some interaction between soil organisms such as competition, mutualism, predation, and parasitism (Bicharanloo et al. 2020; Zhou et al. 2020; Kanté et al. 2021). In this context, we must consider three main laws for root exudation and trophic structure:

- 1st Law – The dynamic: The rhizospheric zone will always be more active and dynamic when compared to the bulk soil.
- 2nd Law – The changes: Root exudates are constantly changing the rhizosphere microbiome.
- 3rd Law – The feedback: Organic acids released by the plant roots influence the ability of plant uptake nutrients and water.

The combination of these three laws directly affects the plant community and its fitness at the tropics (Kuzyakov and Razavi 2019; Li et al. 2020).

## 5.3   Net Primary Production

In soil ecology, primary production is defined as the sum of all organic compounds synthetized by the autotrophs (e.g., algae, bacteria, and vascular and non-vascular plants) through both photosynthesis and chemosynthesis (Eqs. 5.1 and 5.2) from atmospheric carbon dioxide (Gough et al. 2019; Roth et al. 2019; Bay et al. 2021). The autotrophs responsible for these processes are also known as primary producers as described by Souza and Freitas (2018). Also, both equations described further show the production of energy in organic compounds by all living autotrophs (Ignatov 2019; Hippler et al. 2021).

$$\text{Photosynthesis' equation}: CO_2 + H_2O \xrightarrow{\text{light}} CH_2O + O_2 \qquad (5.1)$$

$$\text{Chemosynthesis' equation}: CO_2 + O_2 + 4H_2S \xrightarrow{\text{light}} CH_2O + 4S + 3H_2O \qquad (5.2)$$

In this context, we can find two main concepts: (i) gross primary production and (ii) net primary production. The first one is the amount of chemical energy produced by the primary producers, and it is typically expressed as C biomass (Masciandaro et al. 2018; Anbarashan et al. 2020), whereas the second one is the remaining fixed chemical energy (Eq. 5.3) after some fixed energy be used for maintenance of living tissues and for cellular respiration (Malhotra et al. 2018; Vanlerberghe et al. 2020).

$$\begin{aligned} &\textit{Net} \text{ primary production equation}: \textit{Net} \text{ primary production} = \\ &\text{Gross primary production} - \text{maintenance} - \text{respiration} \qquad (5.3) \end{aligned}$$

The net primary production is also defined as the chemical energy readily available for organism consumption. It is expressed in $gCm^{-2}$, and it is the main source of energy for the following:

(a) Herbivores, by feeding fresh plant tissue (Schaeffer et al. 2018; Frew et al. 2019; Waterman et al. 2019)
(b) Litter transformers, by eating dead plant tissue on soil surface (Sauvadet et al. 2017; Forstall-Sosa et al. 2020; Jo et al. 2020; Rubio-Ríos et al. 2021)
(c) Decomposers, by using the chemical (enzymatic) pathway to decompose the soil organic matter (van der Wal and de Boer 2017; Liu et al. 2018; Wang et al. 2021a)

## 5.4  Plant Litter

Plant litter is defined as the amount of dead plant tissue deposited on soil surface (Ristok et al. 2017; Cao et al. 2020; Tong et al. 2021). On the other hand, some authors have described it as all dead plant material smaller than 2 cm (Cornelissen et al. 2017; Seer et al. 2021). This second definition is useful for scientific studies which consider plant litter as an aleatory variable. This dead plant material may also be called as detritus (Demi et al. 2018; Di Sabatino et al. 2020; Seer et al. 2021), and the dead plant tissues larger than 2 cm are classed as coarse litter (Bani et al. 2018; Zuo et al. 2018). The plant litter is a function of seasonality, ecosystem type, plant diversity, soil fertility, altitude, and plant growth (González-Rodríguez et al. 2018;

Kumar et al. 2018; van der Sande et al. 2018; Nonghuloo et al. 2020; Kitayama et al. 2021).

For soil ecologists, the plant litter layer formed on soil surface constitutes the O horizon which can be classified in three different layers (Li et al. 2017; Ge et al. 2017; Kyaschenko et al. 2017):

1. L-layer: It is the relatively undecomposed plant material deposited on soil surface. The morphological characteristics of this material remain preserved.
2. F-layer: It is characterized by the accumulation of partly decomposed organic matter, commonly described as litter-soil interface.
3. H-layer: It is characterized by the accumulation of fully decomposed organic matter. Here it is impossible to discern the decomposed plant material from soil particles.

It constitutes the main nutrient and energy pool added to the top layer of the soil ecosystem, and it may vary in quantity, decomposition rate, and nutrient content (Xiao et al. 2019; de Almeida et al. 2020; Wang et al. 2021b), and it acts as the main source of habitat and energy supply for many soil organisms such as litter transformers, decomposers, predators, microregulators, and all kind of larvae (Souza and Freitas 2018; Lin et al. 2019; Wu and Wang 2019). Plant litter is an easy bioindicator (e.g., commonly expressed as $gm^{-2}$) to provide information for ecosystem dynamics since it gives evidence about how does the primary production or nutrient cycling work into the soil ecosystem (Chen et al. 2017; Hoosbeek et al. 2018; Osborne et al. 2020). Six positive effects of plant litter are as follows:

1. It facilitates rainwater capture and infiltration.
2. It protects soil aggregates from raindrop impact.
3. It prevents the release of clay and silt particles from plugging soil pores.
4. It decreases cross surface flow.
5. It deaccelerates soil erosion (e.g., wind erosion).
6. It protects soil from wildfire damage.

At the tropics, litter layer is scars due to year-round decomposition, high vegetation variation, seasonality, and plant growth strategies (e.g., some plant species lose their leaves during dry periods) (Giweta 2020; de Oliveira et al. 2020; Kothandaraman et al. 2020; Raj and Jhariya 2021). Plant litter is also directly connected to net primary production (e.g., due to their losses for herbivory and litterfall). Thus, we must consider that the net primary production and litter deposition present similar patterns at both temporal and spatial scales (Costa et al. 2020; Linger et al. 2020). In fact, litter provides habitat for a wide variety of soil organisms (Luan et al. 2021; Lucena et al. 2021). Certain plant species at the tropics are adapted to germinate into the litter layer (Zebaze et al. 2021). Other organisms whose diet consist of any kind of plant tissue (e.g., fresh or dead) live surrounding it, such as algae, bacteria, beetles, centipedes, earthworms, fungi, insect larvae, mollusks, mites, nematodes, protozoa, springtails, and tardigrades (Yang et al. 2018; Gongalsky 2021). The consumption of litter by all these soil organisms results in three different pathways as already described in Chap. 3. They can release organic material and nutrients into the soil profile where the decomposers and plant can use them, thus creating positive plant-soil feedback (Yang and Li 2020; Long et al. 2021).

## 5.5 Conclusions

Trophic structure and soil biological communities are intimately correlated in soil profile. The first one is defined by groups of soil organisms and their function inside the soil food web, while the second one is defined by taxonomical groups. However, both are very influenced by three main complex factors: (i) root exudation, (ii) net primary production, and plant litter. Root exudation creates an active zone which promotes all biochemical reactions. It creates a dynamic biome called rhizosphere. Net primary production describes how healthy is the soil ecosystem through biomass production. This biomass acts as energy source for herbivores, while it also acts as a plant litter pool. Plant litter that is considered as habitat for algae, bacteria, beetles, centipedes, earthworms, fungi, insect larvae, mollusks, mites, nematodes, protozoa, springtails, and tardigrades. We must consider the interaction of these three factors and how do they affect trophic structure and biological community at the tropics.

## References

Anbarashan M, Padmavathy A, Alexandar A, Dhatchanamoorhty N (2020) Survival, growth, aboveground biomass, and carbon sequestration of mono and mixed native tree species plantations on the Coromandel Coast of India. Geol Ecol Landsc 4(2):111–120. https://doi.org/10.1080/24749508.2019.1600910

Arroyo AI, Pueyo Y, Giner ML, Foronda A, Sanchez-Navarrete P, Saiz H, Alados CL (2018) Evidence for chemical interference effect of an allelopathic plant on neighboring plant species: a field study. PLoS One 13(2):e0193421. https://doi.org/10.1371/journal.pone.0193421

Aslam F, Khaliq A, Matloob A, Tanveer A, Hussain S, Zahir ZA (2017) Allelopathy in agroecosystems: a critical review of wheat allelopathy-concepts and implications. Chemoecology 27:1–24. https://doi.org/10.1007/s00049-016-0225-x

Bani A, Pioli S, Ventura M, Panzacchi P, Borruso L, Tognetti R, Tono G, Brusetti L (2018) The role of microbial community in the decomposition of leaf litter and deadwood. Appl Soil Ecol 126:75–84. https://doi.org/10.1016/j.apsoil.2018.02.017

Barbato D, Perini C, Mocali S, Bacaro G, Tordoni E, Maccherini S, Marchi M, Cantiani P, de Meo I, Bianchetto E, Landi S, Bruschini S, Bettini G, Gardin L, Salerni E (2019) Teamwork makes the dream work: disentangling cross-taxon congruence across soil biota in black pine plantations. Sci Total Environ 656:659–669. https://doi.org/10.1016/j.scitotenv.2018.11.320

Bay SK, Waite DW, Dong X, Gillor O, Chown SL, Hugenholtz P, Greening C (2021) Chemosynthetic and photosynthetic bacteria contribute differentially to primary production across a steep desert aridity gradient. ISME J:1–18. https://doi.org/10.1038/s41396-021-01001-0

Bicharanloo B, Shirvan MB, Keitel C, Dijkstra FA (2020) Rhizodeposition mediates the effect of nitrogen and phosphorous availability on microbial carbon use efficiency and turnover rate. Soil Biol Biochem 142:107705. https://doi.org/10.1016/j.soilbio.2020.107705

Bluhm C, Butenschoen O, Maraun M, Scheu S (2019) Effects of root and leaf litter identity and diversity on oribatid mite abundance, species richness and community composition. PLoS One 14(7):e0219166. https://doi.org/10.1371/journal.pone.0219166

Cantó C de la F, Simonin M, King E, Moulin L, Bennett MJ, Castrillo G, Laplaze L (2020) An extended root phenotype: the rhizosphere, its formation and impacts on plant fitness. Plant J 103(3):951–964. https://doi.org/10.1111/tpj.14781

Cao J, He X, Chen Y, Zhang Y, Yu S, Zhou L, Li Z, Zhang C, Fu S (2020) Leaf litter contributes more to soil organic carbon than fine roots in two 10-year-old subtropical plantations. Sci Total Environ 704:135341. https://doi.org/10.1016/j.scitotenv.2019.135341

Capek P, Manzoni S, Kaštovká E, Wild B, Diákova K, Bárta J, Schnecker J, Biasi C, Mastikainen PJ, Alves RJE, Gentsch N, Hugelius G, Plamtag J, Mikutta R, Shibistova O, Urich T, Schleper C, Richte A, Šantůčková H (2018) A plant–microbe interaction framework explaining nutrient effects on primary production. Nat Ecol Evol 2:1588–1596. https://doi.org/10.1038/s41559-018-0662-8

Certini G, Moya D, Lucas-Borja ME, Mastrolonardo G (2021) The impact of fire on soil-dwelling biota: a review. For Ecol Manag 488:118989. https://doi.org/10.1016/j.foreco.2021.118989

Chen HYH, Brant AN, Seedre M, Brassad BW, Taylor AR (2017) The contribution of litterfall to net primary production during secondary succession in the boreal forest. Ecosystems 20:830–844. https://doi.org/10.1007/s10021-016-0063-2

Cornelissen JH, Grootemaat S, Verheijen LM, Cornwell WK, van Bodegom PM, van der Wal R, Aerts R (2017) Are litter decomposition and fire linked through plant species traits? New Phytol 216(3):653–669. https://doi.org/10.1111/nph.14766

Costa AN, Souza JR, Alves KM, Penna-Oliveira A, Paula-Silva G, Becker IS, Marinho-Vieira K, Bonfim AL, Bartimachi A, Vieira-Neto EHM (2020) Linking the spatiotemporal variation of litterfall to standing vegetation biomass in Brazilian savannas. J Plant Ecol Vol 13(5):517–524. https://doi.org/10.1093/jpe/rtaa039

da Silva ER, Overbeck GE, Soares GLG (2017) Something old, something new in allelopathy review: what grassland ecosystems tell us. Chemoecology 27:217–231. https://doi.org/10.1007/s00049-017-0249-x

da Silva SIA, Souza TAF, Lucena EO, da Silva LJR, Laurindo LK, Nascimento GS, Santos D (2021) High phosphorus availability promotes the diversity of arbuscular mycorrhizal spores' community in different tropical crop systems. Biologia. https://doi.org/10.1007/s11756-021-00874-y

Dastogeer KM, Tumpa FH, Sultana A, Akter MA, Chakraborty A (2020) Plant microbiome—an account of the factors that shape community composition and diversity. Curr Plant Biol 23:100161. https://doi.org/10.1016/j.cpb.2020.100161

de Almeida HAD, Ramos MB, Diniz FC, Lopes SDF (2020) What role does elevational variation play in determining the stock and composition of litter? Floresta e Ambiente 27. https://doi.org/10.1590/2179-8087.019618

de Oliveira ACP, Nunes A, Rodrigues RG, Branquinho C (2020) The response of plant functional traits to aridity in a tropical dry forest. Sci Total Environ 747:141177. https://doi.org/10.1016/j.scitotenv.2020.141177

Demi LM, Benstead JP, Rosemond AD, Maerz JC (2018) Litter P content drives consumer production in detritus-based streams spanning an experimental N: P gradient. Ecology 99(2):347–359. https://doi.org/10.1002/ecy.2118

Di Sabatino A, Cicolani B, Miccoli FP, Cristiano G (2020) Plant detritus origin and microbial–detritivore interactions affect leaf litter breakdown in a Central Apennine (Italy) cold spring. Aquat Ecol 54:495–504. https://doi.org/10.1007/s10452-020-09755-z

Eisenhauer N, Lanoue A, Strecker T, Scheu S, Steinauer K, Thakur MP, Mommer L (2017) Root biomass and exudates link plant diversity with soil bacterial and fungal biomass. Sci Rep 7:44641. https://doi.org/10.1038/srep44641

Forstall-Sosa KS, Souza TAF, Lucena EO, Silva SIA, Ferreira JTA, Silva TN, Ferreira JTA, Silva TN, Santos D, Niemeyer JC (2020) Soil macroarthropod community and soil biological quality index in a green manure farming system of the Brazilian semi-arid. Biologia. https://doi.org/10.2478/s11756-020-00602-y

Frew A, Weston LA, Gurr GM (2019) Silicon reduces herbivore performance via different mechanisms, depending on host–plant species. Austral Ecol 44(6):1092–1097. https://doi.org/10.1111/aec.12767

Ge X, Xiao W, Zeng L, Huang Z, Zhou B, Schaub M, Li MH (2017) Relationships between soil–litter interface enzyme activities and decomposition in Pinus massoniana plantations in China. J Soils Sediments 17(4):996–1008. https://doi.org/10.1007/s11368-016-1591-2

Giweta M (2020) Role of litter production and its decomposition, and factors affecting the processes in a tropical forest ecosystem: a review. J Ecol Environ 44:11. https://doi.org/10.1186/s41610-020-0151-2

Glatthorn J, Annigöfer P, Balkenhol N, Leuschner C, Polle A, Scheu S, Schuldt A, Schildt B, Ammer C (2021) An interdisciplinary framework to describe and evaluate the functioning of forest ecosystems. Basic Appl Ecol 52:1–14. https://doi.org/10.1016/j.baae.2021.02.006

Gongalsky KB (2021) Soil macrofauna: study problems and perspectives. Soil Biol Biochem 159:108281. https://doi.org/10.1016/j.soilbio.2021.108281

González-Rodríguez H, Ramírez-Lozano RG, Cantú-Silva I, Gómez-Meza MV, Estrada-Castillón E, Arévalo JR (2018) Deposition of litter and nutrients in leaves and twigs in different plant communities of northeastern Mexico. J For Res 29(5):1307–1314. https://doi.org/10.1007/s11676-017-0553-x

Gough CM, Atkins JW, Fahey RT, Hardiman BS (2019) High rates of primary production in structurally complex forests. Ecology 100(10):e02864. https://doi.org/10.1002/ecy.2864

Guo Y, Hou L, Zhang Z, Zhang J, Cheng J, Wei G, Lin Y (2019) Soil microbial diversity during 30 years of grassland restoration on the Loess Plateau, China: tight linkages with plant diversity. Land Degrad Dev 30(10):1172–1182. https://doi.org/10.1002/ldr.3300

Henneron L, Cros C, Picon-Cochard C, Rahimian V, Fontaine S (2020) Plant economic strategies of grassland species control soil carbon dynamics through rhizodeposition. J Ecol 108(2):528–545. https://doi.org/10.1111/1365-2745.13276

Herz K, Dietz S, Gorzolka K, Haider S, Jandt U, Scheel D, Bruelheide H (2019) Linking root exudates to functional plant traits. PLoS One 14(3):e0213965. https://doi.org/10.1371/journal.pone.0213965

Hickman DT, Rasmussen A, Ritz K, Birkett MA, Neve P (2021) Allelochemicals as multi-kingdom plant defence compounds: towards an integrated approach. Pest Manag Sci 77(3):1121–1131. https://doi.org/10.1002/ps.6076

Hippler M, Minagawa J, Takahashi Y (2021) Photosynthesis and chloroplast regulation—balancing photosynthesis and photoprotection under changing environments. Plant Cell Physiol 62(7):1059–1062. https://doi.org/10.1093/pcp/pcab139

Hoosbeek MR, Remme RP, Rusch GM (2018) Trees enhance soil carbon sequestration and nutrient cycling in a silvopastoral system in south-western Nicaragua. Agrofor Syst 92:263–273. https://doi.org/10.1007/s10457-016-0049-2

Hupe A, Schulz H, Bruns C, Haase T, Heß J, Joergensen RG, Wichern F (2018) Even flow? Changes of carbon and nitrogen release from pea roots over time. Plant Soil 431(1):143–157. https://doi.org/10.1007/s11104-018-3753-z

Hussain MI, Danish S, Sánchez-Moreiras AM, Vicente Ó, Jabran K, Chaudhry UK, Branca F, Reigosa MJ (2021) Unraveling sorghum allelopathy in agriculture: concepts and implications. Plan Theory 10(9):1795. https://doi.org/10.3390/plants10091795

Igiehon NO, Babalola OO (2020) Rhizosphere microbiome modulators: contributions of nitrogen fixing bacteria towards sustainable agriculture. Int J Environ Res Public Health 15(4):574. https://doi.org/10.3390/ijerph15040574

Ignatov I (2019) Origin of life in hot mineral water from hydrothermal springs and ponds. Effects of hydrogen and nascent hydrogen. Analyses with spectral methods, pH and ORP. Eur Rev Chem Res 6(2):49–60. https://doi.org/10.13187/ercr.2019.2.49

Jackson LE, Bowles TM, Ferris H, Margenot AJ, Hollander A, Garcia-Palacios P, Daufresne T, Sánchez-Moreno S (2019) Plant and soil microfaunal biodiversity across the borders between arable and forest ecosystems in a Mediterranean landscape. Soil Appl Ecol 136:122–138. https://doi.org/10.1016/j.apsoil.2018.11.015

Jeewani PH, Gunina A, Tao L, Zhu Z, Kuzyakov Y, Van Zwieten L, Guggenberger G, Shen C, Yu G, Singh BP, Pan S, Luo Y, Xu J (2020) Rusty sink of rhizodeposits and associated keystone microbiomes. Soil Biol Biochem 147:107840. https://doi.org/10.1016/j.soilbio.2020.107840

Jo I, Fridley JD, Frank DA (2020) Rapid leaf litter decomposition of deciduous understory shrubs and lianas mediated by mesofauna. Plant Ecol 221:63–68. https://doi.org/10.1007/s11258-019-00992-3

Kanté M, Riah-Anglet W, Cliquet J-B, Trinsoutrot-Gattin I (2021) Soil enzyme activity and stoichiometry: linking soil microorganism resource requirement and legume carbon rhizodeposition. Agronomy 11(11):2131. https://doi.org/10.3390/agronomy11112131

Kitayama K, Ushio M, Aiba SI (2021) Temperature is a dominant driver of distinct annual seasonality of leaf litter production of equatorial tropical rain forests. J Ecol 109(2):727–736. https://doi.org/10.1111/1365-2745.13500

Klimešová J, Martínkova J, Ottaviani G (2018) Belowground plant functional ecology: towards an integrated perspective. Fuct Ecol 32(9):2115–2126. https://doi.org/10.1111/1365-2435.13145

Kothandaraman S, Dar JA, Sundarapandian S, Dayanandan S, Khan ML (2020) Ecosystem-level carbon storage and its links to diversity, structural and environmental drivers in tropical forests of Western Ghats, India. Sci Rep 10(1):1–15. https://doi.org/10.1038/s41598-020-70313-6

Kumar P, Chen HYH, Thomas CS, Shahi C (2018) Linking resource availability and heterogeneity to understorey species diversity through succession in boreal forest of Canada. J Ecol 106(3):1266–1276. https://doi.org/10.1111/1365-2745.12861

Kuzyakov Y, Razavi BS (2019) Rhizosphere size and shape: temporal dynamics and spatial stationarity. Soil Biol Biochem 135:343–360. https://doi.org/10.1016/j.soilbio.2019.05.011

Kyaschenko J, Clemmensen KE, Karltun E, Lindahl BD (2017) Below-ground organic matter accumulation along a boreal forest fertility gradient relates to guild interaction within fungal communities. Ecol Lett 20(12):1546–1555. https://doi.org/10.1111/ele.12862

Latif S, Chiapusio G, Weston LA (2017) Allelopathy and the Role of Allelochemicals in Plant Defence. In: Advances in Botanical Research, Vol. 82, Academic Press, Cambridge, 19-54. https://doi.org/10.1016/bs.abr.2016.12.001

Legrand F, Picot A, Cobo-Diaz JF, Carof M, Chen W, Le Floch G (2018) Effect of tillage and static abiotic soil properties on microbial diversity. Appl Soil Ecol 132:135–145. https://doi.org/10.1016/j.apsoil.2018.08.016

Li X, Xiao Q, Niu J, Dymond S, McPherson EG, van Doorn N, Yu X, Xie B, Zhang K, Li J (2017) Rainfall interception by tree crown and leaf litter: an interactive process. Hydrol Process 31(20):3533–3542. https://doi.org/10.1002/hyp.11275

Li J, Yuan X, Ge L, Li Q, Li Z, Wang L, Liu Y (2020) Rhizosphere effects promote soil aggregate stability and associated organic carbon sequestration in rocky areas of desertification. Agric Ecosyst Environ 304:107126. https://doi.org/10.1016/j.agee.2020.107126

Lin D, Wang F, Fanin N, Pang M, Dou P, Wang H, Qian S, Zhao L, Yang Y, Mi X, Ma K (2019) Soil fauna promote litter decomposition but do not alter the relationship between leaf economics spectrum and litter decomposability. Soil Biol Biochem 136:107519. https://doi.org/10.1016/j.soilbio.2019.107519

Linger E, Hogan JA, Cao M, Zhang WF, Yang XF, Hu YH (2020) Precipitation influences on the net primary productivity of a tropical seasonal rainforest in Southwest China: a 9-year case study. For Ecol Manag 467:118153. https://doi.org/10.1016/j.foreco.2020.118153

Liu W, Qiao C, Yang S, Bai W, Liu L (2018) Microbial carbon use efficiency and priming effect regulate soil carbon storage under nitrogen deposition by slowing soil organic matter decomposition. Geoderma 332:37–44. https://doi.org/10.1016/j.geoderma.2018.07.008

Liu Z, Sun Y, Zhang Y, Feng W, Lai Z, Qin S (2021) Soil microbes transform inorganic carbon into organic carbon by dark fixation pathways in desert soil. Journal of geophysical research. Biogeosciences:e2020JG006047. https://doi.org/10.1029/2020JG006047

Long J, Zhang M, Li J, Liao H, Wang X (2021) Soil macro-and mesofauna-mediated litter decomposition in a subtropical karst forest. Biotropica 53(6):1465–1474. https://doi.org/10.1111/btp.12980

Luan J, Liu S, Li S, Whalen JK, Wang Y, Wang J, Liu Y, Dong W, Chang SX (2021) Functional diversity of decomposers modulates litter decomposition affected by plant invasion along a climate gradient. J Ecol 109(3):1236–1249. https://doi.org/10.1111/1365-2745.13548

Lucena EO, Souza TAF, da Silva SIA, Kormann S, da Silva LJR, Laurindo LK, Forstall-Sosa KS, de Andrade LA (2021) Soil biota community composition as affected by Cryptostegia madagascariensis invasion in a tropical Cambisol from North-eastern Brazil. Trop Ecol 62:662–669. https://doi.org/10.1007/s42965-021-00177-y

Mahmud K, Makaju S, Ibrahim R, Missaoui A (2020) Current progress in nitrogen fixing plants and microbiome research. Plan Theory 9(1):97. https://doi.org/10.3390/plants9010097

Mahmud K, Missaoui A, Lee KC, Ghimire B, Presley HW, Makaju S (2021) Rhizosphere microbiome manipulation for sustainable crop production. Curr Plant Biol 27:100210. https://doi.org/10.1016/j.cpb.2021.100210

Malhotra H, Sharma SV, Pandey R (2018) Phosphorus nutrition: plant growth in response to deficiency and excess. In: Hasanuzzaman M, Fujita M, Oku H, Nahar K, Hawrylak-Nowak B (eds) Plant nutrients and abiotic stress tolerance. Springer, Singapore. https://doi.org/10.1007/978-981-10-9044-8_7

Marastoni L, Lucini L, Miras-Moreno B, Trevisan M, Sega D, Zamboni A, Varanini Z (2020) Changes in physiological activities and root exudation profile of two grapevine rootstocks reveal common and specific strategies for Fe acquisition. Sci Rep 10(1):1–12. https://doi.org/10.1038/s41598-020-75317-w

Marian F, Sandmann D, Krashevska V, Maraun M, Scheu S (2018) Altitude and decomposition stage rather than litter origin structure soil microarthropod communities in tropical montane rainforests. Soil Biol Biochem 125:263–274. https://doi.org/10.1016/j.soilbio.2018.07.017

Masciandaro G, Macci C, Peruzzi E, Doni S (2018) Soil carbon in the world: ecosystem services linked to soil carbon in forest and agricultural soils. In: Garcia C, Nannipieri P, Hernandez T (eds) The future of soil carbon. Elsevier, pp 1–38. https://doi.org/10.1016/B978-0-12-811687-6.00001-8

Mehmood A, Hussain A, Irshad M, Hamayun M, Iqbal A, Rahman H, Tawab A, Ahmad A, Ayaz S (2019) Cinnamic acid as an inhibitor of growth, flavonoids exudation and endophytic fungus colonization in maize root. Plant Physiol Biochem 135:61–68. https://doi.org/10.1016/j.plaphy.2018.11.029

Mishra S, Hättenschwiler S, Yang X (2020) The plant microbiome: a missing link for the understanding of community dynamics and multifunctionality in forest ecosystems. Appl Soil Ecol 145:103345. https://doi.org/10.1016/j.apsoil.2019.08.007

Morgan RB, Herrmann V, Kunert N, Lamberty BB, Muller-Landau HC, Anderson-Teixeira KJ (2021) Global patterns of forest autotrophic carbon fluxes. Glob Chang Biol 27(12):2840–2855. https://doi.org/10.1111/gcb.15574

Nielsen SN, Müller F, Marques JC, Bastianoni S, Jørgensen SE (2020) Thermodynamics in ecology – an introductory review. Entropia 22(8):820. https://doi.org/10.3390/e22080820

Nonghuloo IM, Kharbhih S, Suchiang BR, Adhikari D, Upadhaya K, Barik SK (2020) Production, decomposition and nutrient contents of litter in subtropical broadleaved forest surpass those in coniferous forest, Meghalaya. Trop Ecol 61:5–12. https://doi.org/10.1007/s42965-020-00065-x

Novo LAB, Castro PML, Alvarenga P, da Silva EF (2018) Plant growth–promoting rhizobacteria-assisted phytoremediation of mine soils. In: MNV P, PJC F, Maiti SK (eds) Bio-geotechnologies for mine site rehabilitation, pp 281–295. https://doi.org/10.1016/B978-0-12-812986-9.00016-6

Oburger E, Jones DL (2018) Sampling root exudates – Mission impossible? Rhizosphere 6:116–133. https://doi.org/10.1016/j.rhisph.2018.06.004

Olanrewaju OS, Ayangbenro AS, Glick BR, Babalola OO (2019) Plant health: feedback effect of root exudates-rhizobiome interactions. Appl Microbiol Biotechnol 103:1155–1166. https://doi.org/10.1007/s00253-018-9556-6

Osborne BB, Nasto MK, Soper FM, Asner GP, Balzotti CS, Cleveland CC, Taylor PG, Townsend AR, Porder S (2020) Leaf litter inputs reinforce islands of nitrogen fertility in a lowland tropical forest. Biogeochemistry 147(3):293–306. https://doi.org/10.1007/s10533-020-00643-0

Potatov AM, Tiunov AV, Scheu S (2018) Uncovering trophic positions and food resources of soil animals using bulk natural stable isotope composition. Biol Rev 94:37–59. https://doi.org/10.1111/brv.12434

Potatov AM, Rozanova OL, Semenina EE, Leonov VD, Belyakova OI, BogatyrevaVY DMI, Esaulov AS, Korotjevih AY, Kudrin AA, Malysheva EA, Mazei YA, Tsuikov SM, Zuev AG, Tiunov AV (2021) Size compartmentalization of energy channeling in terrestrial belowground food webs. Ecol Soc Am 102(8):e03421. https://doi.org/10.1002/ecy.3421

Preece C, Peñuelas J (2020) A return to the wild: root exudates and food security. Trends Plant Sci 25(1):14–21. https://doi.org/10.1016/j.tplants.2019.09.010

Raj A, Jhariya MK (2021) Site quality and vegetation biomass in the tropical Sal mixed deciduous forest of Central India. Landscape Ecol Eng 17:387–399. https://doi.org/10.1007/s11355-021-00450-1

Rilfe SA, Griffiths J, Ton J (2019) Crying out for help with root exudates: adaptive mechanisms by which stressed plants assemble health-promoting soil microbiomes. Curr Opin Microbiol 49:73–82. https://doi.org/10.1016/j.mib.2019.10.003

Ristok C, Leppert KN, Franke K, Schemer-Lorenzen M, Niklaus PA, Wessjohann LA, Bruelheide H (2017) Leaf litter diversity positively affects the decomposition of plant polyphenols. Plant Soil 419:305–317. https://doi.org/10.1007/s11104-017-3340-8

Robles-Aguilar AA, Schrey SD, Postma JA, Temperton VM, Jablonowski ND (2020) Phosphorus uptake from struvite is modulated by the nitrogen form applied. J Plant Nutr Soil Sci 183(1):80–90. https://doi.org/10.1002/jpln.201900109

Roth MS, Westcott DJ, Iwai M, Niyogi KK (2019) Hexokinase is necessary for glucose-mediated photosynthesis repression and lipid accumulation in a green alga. Commun Biol 2:347. https://doi.org/10.1038/s42003-019-0577-1

Rubio-Ríos J, Pérez J, Salinas MJ, Fenoy E, López-Rojo N, Boyero L, Casas JJ (2021) Key plant species and detritivores drive diversity effects on instream leaf litter decomposition more than functional diversity: a microcosm study. Sci Total Environ 798:149266. https://doi.org/10.1016/j.scitotenv.2021.149266

Sauvadet M, Chauvat M, Brunet N, Bertrand I (2017) Can changes in litter quality drive soil fauna structure and functions? Soil Biol Biochem 107:94–103. https://doi.org/10.1016/j.soilbio.2016.12.018

Scavo A, Abbate C, Mauromicale G (2019) Plant allelochemicals: agronomic, nutritional and ecological relevance in the soil system. Plant Soil 442:23–48. https://doi.org/10.1007/s11104-019-04190-y

Schaeffer RN, Wang Z, Thornber CS, Preisser EL, Orians CM (2018) Two invasive herbivores on a shared host: patterns and consequences of phytohormone induction. Oecologia 186:973–982. https://doi.org/10.1007/s00442-018-4063-0

Seer FK, Putze G, Pennings SC, Zimmer M (2021) Drivers of litter mass loss and faunal composition of detritus patches change over time. Ecol Evol 11(14):9642–9651. https://doi.org/10.1002/ece3.7787

Semchenko M, Saar S, Lepik A (2017) Intraspecific genetic diversity modulates plant–soil feedback and nutrient cycling. New Phytol 216(1):90–98. https://doi.org/10.1111/nph.14653

Sharma A, Verma RK (2018) Root-microbe interactions: understanding and exploitation of microbiome. In: Giri B, Prasad R, Varma A (eds) Root biology. Soil biology, 52. Springer, Cham. https://doi.org/10.1007/978-3-319-75910-4_13

Shen N, Jing Y, Tu G, Fu A, Lan W (2020) Danger-associated peptide regulates root growth by promoting protons extrusion in an AHA2-dependent manner in Arabidopsis. Int J Mol Sci 21(21):7963. https://doi.org/10.3390/ijms21217963

Siao W, Coskun D, Baluška F, Kronzucker HJ, Xu W (2020) Root-apex proton fluxes at the centre of soil-stress acclimation. Trends Plant Sci 25(8):794–804. https://doi.org/10.1016/j.tplants.2020.03.002

Souza TAF, Freitas H (2018) Long-term effects of fertilization on soil organism diversity. In: Gaba S, Smith B, Lichtfouse E (eds) Sustainable agriculture reviews. Springer, Cham, pp 211–247. https://doi.org/10.1007/978-3-319-90309-5_7

Souza TAF, Doubner M Jr, Schmitt DE, da Silva LJR, Shneider K (2022) Soil biotic and abiotic traits as driven factors for site quality of *Araucaria angustifolia* plantations. Biologia. https://doi.org/10.21203/rs.3.rs-445199/v1

Stevens BM, Propster JR, Öpik M, Wilson GWT, Alloway SL, Mayemba E, Johnson NC (2020) Arbuscular mycorrhizal fungi in roots and soil respond differently to biotic and abiotic factors in the Serengeti. Mycorrhiza 30:79–95. https://doi.org/10.1007/s00572-020-00931-5

Sun H, Jiang S, Jiang C, Wu C, Gao M, Wang Q (2021) A review of root exudates and rhizosphere microbiome for crop production. Environ Sci Pollut Res 28:54497–54510. https://doi.org/10.1007/s11356-021-15838-7

Tang X, Pei X, Lei N, Luo X, Liu L, Shi L, Chen G, Liang J (2020) Global patterns of soil autotrophic respiration and its relation to climate, soil and vegetation characteristics. Geoderma 369:114339. https://doi.org/10.1016/j.geoderma.2020.114339

Tang X, Shi Y, Luo X, Liu L, Jian J, Bond-Lamberty B, Hao D, Olchev A, Zhang W, Gao S, Li J (2021) Uma alocação decrescente de carbono para a respiração autotrófica subterrânea em ecossistemas florestais globais. Sci Total Environ 798:149273. https://doi.org/10.1016/j.scitotenv.2021.149273

Tong R, Zhou B, Jiang L, Ge X, Cao Y, Shi J (2021) Leaf litter carbon, nitrogen and phosphorus stoichiometry of Chinese fir (Cunninghamia lanceolata) across China. Glob Ecol Conserv 27:e01542. https://doi.org/10.1016/j.gecco.2021.e01542

Trivedi P, Leach JE, Tringe SG, Sa T, Singh BK (2020) Plant–microbiome interactions: from community assembly to plant health. Nat Rev Microbiol 18:607–621. https://doi.org/10.1038/s41579-020-0412-1

Tsunoda T, van Dam NM (2017) Root chemical traits and their roles in belowground biotic interactions. Pedobiologia 65:58–67. https://doi.org/10.1016/j.pedobi.2017.05.007

Tsurikov SM, Ermilov SG, Tiunov AV (2019) Trophic structure of a tropical soil- and litter-dwelling oribatid mite community and consistency of trophic niches across biomes. Exp Appl Acarol 78:29–48. https://doi.org/10.1007/s10493-019-00374-4

van der Sande MT, Arets EJ, Peña-Claros M, Hoosbeek MR, Cáceres-Siani Y, van der Hout P, Poorter L (2018) Soil fertility and species traits, but not diversity, drive productivity and biomass stocks in a Guyanese tropical rainforest. Funct Ecol 32(2):461–474. https://doi.org/10.1111/1365-2435.12968

van der Wal A, de Boer W (2017) Dinner in the dark: illuminating drivers of soil organic matter decomposition. Soil Biol Biochem 105:45–48. https://doi.org/10.1016/j.soilbio.2016.11.006

Vanlerberghe GC, Dahal K, Alber NA, Chadee A (2020) Photosynthesis, respiration and growth: a carbon and energy balancing act for alternative oxidase. Mitochondrion 52:197–211. https://doi.org/10.1016/j.mito.2020.04.001

Vives-Peris V, de Ollas C, Gómez-Cadenas A, Pérez-Clemente (2020) Root exudates: from plant to rhizosphere and beyond. Plant Cell Rep 39:3–17. https://doi.org/10.1007/s00299-019-02447-5

Wang C, Xue L, Dong Y, Jiao R (2021a) Soil organic carbon fractions, C-cycling hydrolytic enzymes, and microbial carbon metabolism in Chinese fir plantations. Sci Total Environ 758:143695. https://doi.org/10.1016/j.scitotenv.2020.143695

Wang W, Hu K, Huang K, Tao J (2021b) Mechanical fragmentation of leaf litter by fine root growth contributes greatly to the early decomposition of leaf litter. Glob Ecol Conserv 26:e01456. https://doi.org/10.1016/j.gecco.2021.e01456

Waterman JM, Cazzonelli CI, Hartley SE, Johnson SN (2019) Simulated herbivory: the key to disentangling plant defence responses. Trends Ecol Evol 34(5):447–458. https://doi.org/10.1016/j.tree.2019.01.008

Wu P, Wang C (2019) Differences in spatiotemporal dynamics between soil macrofauna and mesofauna communities in forest ecosystems: the significance for soil fauna diversity monitoring. Geoderma 337:266–272. https://doi.org/10.1016/j.geoderma.2018.09.031

Xia Z, Yu L, He Y, Korpelainnen H, Li C (2019) Broadleaf trees mediate chemically the growth of Chinese fir through root exudates. Biol Fertil Soils 55:737–749. https://doi.org/10.1007/s00374-019-01389-0

Xiao W, Chen HY, Kumar P, Chen C, Guan Q (2019) Multiple interactions between tree composition and diversity and microbial diversity underly litter decomposition. Geoderma 341:161–171. https://doi.org/10.1016/j.geoderma.2019.01.045

Yang X, Li T (2020) Effects of terrestrial isopods on soil nutrients during litter decomposition. Geoderma 376:114546. https://doi.org/10.1016/j.geoderma.2020.114546

Yang G, Wagg C, Veresoglou SD, Hempel S, Rillig MC (2018) How soil biota drive ecosystem stability. Trends Plant Sci 23(12):1057–1067. https://doi.org/10.1016/j.tplants.2018.09.007

Zebaze D, Fayolle A, Daïnou K, Libalah M, Droissart V, Sonké B, Doucet JL (2021) Land use has little influence on the soil seed bank in a central African moist forest. Biotropica. https://doi.org/10.1111/btp.13032

Zhang R, Vivanco JM, Shen Q (2017) The unseen rhizosphere root–soil–microbe interactions for crop production. Curr Opin Microbiol 37:8–14. https://doi.org/10.1016/j.mib.2017.03.008

Zhang X, Zhang D, Sun W, Wang T (2019) The adaptive mechanism of plants to iron deficiency via iron uptake, transport, and homeostasis. Int J Mol Sci 20(10):2424. https://doi.org/10.3390/ijms20102424

Zhang Z, Liu Y, Yuan L, Weber E, van Kleunen M (2021a) Effect of allelopathy on plant performance: a meta-analysis. Ecol Lett 24(2):348–362. https://doi.org/10.1111/ele.13627

Zhang K, Maltais-Landry G, Liao HL (2021b) How soil biota regulate C cycling and soil C pools in diversified crop rotations. Soil Biol Biochem:108219. https://doi.org/10.1016/j.soilbio.2021.108219

Zhao Q, Tang J, Li Z, Yang W, Duan Y (2018) The influence of soil physico-chemical properties and enzyme activities on soil quality of saline-alkali agroecosystems in Western Jilin Province, China. Sustainability 10(5):1529. https://doi.org/10.3390/su10051529

Zhao M, Zhao J, Yuan J, Hale L, Wen T, Huang Q, Vivanco JM, Zhou J, Kowalchuk GA, Shen Q (2021) Root exudates drive soil-microbe-nutrient feedbacks in response to plant growth. Plant Cell Environ 44(2):613–628. https://doi.org/10.1111/pce.13928

Zhengtao Z, Wenge H, Di Y, Yuan H, Tingting Z (2019) Diversity of Azotobacter in relation to soil environment in Ebinur Lake wetland. Biotechnol Biotechnol Equip 33:1280–1290. https://doi.org/10.1080/13102818.2019.1659181

Zhou J, Zang H, Loeppmann S, Gube M, Kuzyakov Y, Pausch J (2020) Arbuscular mycorrhiza enhances rhizodeposition and reduces the rhizosphere priming effect on the decomposition of soil organic matter. Soil Biol Biochem 140:107641. https://doi.org/10.1016/j.soilbio.2019.107641

Zuo J, Hefting MM, Berg MP, van Logtestijn RS, van Hal J, Goudzwaard L, Liu J, Sass-Klaassen U, Sterck FJ, Poorter L, Cornelissen JH (2018) Is there a tree economics spectrum of decomposability? Soil Biol Biochem 119:135–142. https://doi.org/10.1016/j.soilbio.2018.01.019

# Chapter 6
# Land Use and Soil Contamination in Dry Tropical Ecosystems

**Abstract** This chapter explores how land uses and soil contamination in dry tropical ecosystems must affect soil organisms' community composition. Land use is defined by the complex of vegetation type, soil management, and aftercare practices commonly used in agroecosystems, while the soil contamination defines the input level of exotic compounds in soil ecosystem that must affect organism' fitness, reproduction, and behavior. These two concepts are strongly linked to each other in dry tropical ecosystem, and in some cases, they interact in the production with negative effects on the entire soil food web. Conventional farming system can negatively influence soil food web overtime by reducing soil organic matter contents. On the other hand, organic farming systems may improve soil organisms' community composition by improving both habitat and resources availability. It has been recognized that organic residues are important drivers of soil biological communities. This chapter examines some of the many ways that soil contamination can be studied by using soil organisms as bioindicators.

**Keywords** Agroecosystems · Conventional farming systems · Degraded ecosystems · Organic farming systems · Soil food web

**Questions Covered in the Chapter**
1. Why does the continuous use of organic fertilizer become more benefic to soil ecosystem than the use of mineral fertilizer in the same tropical conditions?
2. Why is it important to consider the input of organic residues into a dry tropical ecosystem?
3. What is soil contamination?
4. How can we study soil contamination by using soil organisms as bioindicators?

## 6.1   Introduction

In tropical ecosystems, the production of food, fiber, and energy is based on different land uses (e.g., conventional farming systems, organic farming systems, agroforestry system, etc.) that can promote directly or indirectly any degree of
contamination into soil profile (Thangavel and Sridevi 2017; Morgado et al. 2018;
Sanaullah et al. 2020). Each land use has their specific inputs of external chemical
or organic compounds based on a soil management schedule (Schrama et al. 2018;
van der Bom et al. 2018). In this chapter, our focus will be on mineral fertilizations
and the use of some *icide*-type products (e.g., herbicides, rodenticides, insecticides,
and fungicides). The use of these products has been a historical contribution to the
crop system development since the 1950s (Souza et al. 2018). Nowadays, they are
recognized as the main drivers for soil contamination and biodiversity loss (Daam
et al. 2019; Geisen et al. 2019; Sofo et al. 2020). Also, we must consider their negative and positive influences on soil chemical, physical, and biological properties
(Lykogianni et al. 2021; Sánchez-Bayo 2021).

There are some studies showing strong evidence that fertilization can affect soil
biota community structure (Zhao et al. 2019; Ikoyi et al. 2020; Forstall-Sosa et al.
2020; Puissant et al. 2021). The function of these soil organisms is very compromised too (Wang et al. 2018; Menta and Remelli 2020) because soil organisms are
C and N dependent into soil profile. While N and P contents into soil profile generally limit crop production, soil biota is C and N limited as described in detail by
Souza and Freitas (2018). Land use, soil contamination, or both can affect aboveground community, which in turn changes the diversity, activity, and function of the
belowground community (Siebert et al. 2020; Lecerf et al. 2021). Increases in N or
P inputs through mineral fertilization affect negatively arbuscular mycorrhizal community and other soil symbionts. Conversely, the organic fertilization after a long-
term utilization can improve soil biota activity by improving energy supply via soil
organic matter and habitat by promoting soil structure and profile (Lin et al. 2019;
Aldebron et al. 2020; Menta et al. 2020).

Thus, both the aboveground and belowground fitness depend on the land use
(Olorunfemi et al. 2019; Yadav et al. 2019). Here, depending on the kind of fertilization practice (e.g., mineral vs. organic), we must find negative or positive influences
on soil biota response. On the other hand, always these two groups will be negatively influenced by soil contamination (Feng et al. 2018; Tang et al. 2019). Here,
we can find linear, exponential, or log decreases on plant growth, soil biota diversity, and soil organisms' fitness and reproduction (DalCorso et al. 2019; Feng et al.
2022). Other interesting soil organism behavior is the avoidance in contaminated
environments. In such cases, we find a significant increase in soil biota avoidance in
disturbed environments with the large use of toxic compounds (Gomes et al. 2017;
Renaud et al. 2018).

Considering the conventional farming systems as the major agricultural practice
into the tropical zone, we must consider several problems related to the long-term
use of mineral fertilizers (Table 6.1) that can become extremely dangerous for our

**Table 6.1** Main effects of long-term conventional farming system on ecosystem services at the tropics

| Ecosystem services | Main cause | Main effect | Soil biota compromised |
|---|---|---|---|
| Primary production | Decline in both nutrient cycling and soil organic matter dynamics Soil structure loss | Less primary production Less litter deposition Decrease soil quality | Ecosystem engineers Predators Decomposers Microregulators Symbionts |
| Water supply | Soil structure loss Decrease water quality | Increase soil erosion Decrease soil organic matter Increase N and P leaching | Ecosystem engineers Decomposers |
| Climate change | Carbon dynamic Atmospheric $CO_2$ concentration | Accelerate climate change Increase $NO_2$ e $CO_2$ emission | Decomposers |
| Erosion control | Soil structure loss | Increase soil erosion | Decomposers Predators Symbionts Microregulators |
| Soil contamination | Increase soil heavy metal contents Changes into soil food web | Increase soil pollutants Functional redundancy is more frequent | Decomposers Symbionts Microregulators |
| Disease and pest control | Decrease plant resistance Decrease in phenotypic plasticity Increase plant mortality | Changes in soil pH Decrease in soil organic matter Decrease in soil fertility | Predators Microregulators Symbionts |
| Biodiversity conservation | Habitat loss Decrease soil health Increase soil disturbance | Soil structure loss Litter loss Increase soil surface exposure | All soil food web |

Adapted from Souza and Freitas (2018)

survival, for soil ecology, and for the environmental health (Singh 2018; Yang et al. 2020). In fact, conventional farming system affects soil organisms' structure (Harkes et al. 2019; Dangi et al. 2020). Over time, this practice may reduce primary production, water supply, and soil biodiversity as described below. These reductions occur mainly through leaching, soil erosion, crop harvesting, soil contamination, climate change, and low C-input systems (Liu et al. 2018; Oshunsanya et al. 2018; Figlioli et al. 2019; Alewell et al. 2020). The loss on soil structure, habitat, soil organic matter, and nutrient cycling are the main factors contributing to the soil quality decline in tropical soil where there is a long-term use of conventional farming systems (Franco et al. 2020; Thomaz and Antoneli 2021).

In fact, land use and soil contamination have a substantial and ever-growing influence on soil ecosystem to the extent that much of the tropical zone has been totally transformed by a suite of human activities (Oumenskou et al. 2018; Martins et al. 2019). The most evident influence of land use or soil contamination is through the increasing use of fertilizer and *icide*-type products (Hossain et al. 2020; Alvarenga et al. 2020), which at present represents the main driving force for biodiversity, water quality, and soil quality loss (Tsai 2019; Ramirez-Morales et al. 2021). It has been argued that to understand impacts of land use and soil contamination on soil ecosystem and the pathways which underlie them requires explicit consideration of trade-offs between aboveground and belowground diversity. This chapter explores this issue, discussing how particular land use (conventional vs. organic farming systems) and soil contamination impact on soil ecosystem, soil biota diversity, and function.

## 6.2   Land Use and Its Effects on Soil Biological Properties

Land use practices that can increase soil fertility in tropical agroecosystem or agro-forestry systems through the continuous input of essential plant nutrients (e.g., N, P, K, Ca, Mg, S, and micronutrients) are the main pathway to influence soil biological properties (e.g., abundance, diversity, and function) (Siebert et al. 2019; Ikoyi et al. 2020). In a short timescale, the conventional farming system can improve soil functioning and increase plant yield (Brunetti et al. 2019; Ning et al. 2020). However, in a long timescale, conventional farming system with mineral fertilization and the use of *icide*-type products creates negative plant-soil feedback by reducing C input (Purwanto and Alam 2020; Medorio-García et al. 2020), increasing C losses, and in some cases promoting N and P losses through soil erosion (Alewell et al. 2020; Huang et al. 2020). In this context, organic farming system and agroforestry systems which promote slowly increases in soil fertility can create positive plant-soil feedback by promoting soil food web and reducing soil erosion (Lori et al. 2017; Muchane et al. 2020).

Although conventional farming system could provide essential plant nutrient by inorganic fertilization, it could lead to decrease soil quality by reducing soil organic carbon and soil biota diversity (Ozlu and Kumar 2018; Raza et al. 2020). On the other hand, organic farming system leads to increase soil organic carbon and essential nutrients (e.g., N, P, and K) for plant species and soil biota (Barbosa et al. 2021; Nascimento et al. 2021). Figure 6.1 illustrates the long-term effects of both conventional and organic farming system on soil organic carbon, total nitrogen, and soil biota diversity (presented by Shannon's diversity index).

These results showed the positive effect of organic farming system over time on shot biomass, litter deposition, nutrient contents, soil fauna community composition, and microbial biomass (Martínez-García et al. 2018; Li et al. 2021a; Yang et al. 2021) and could have been attributed to the following:

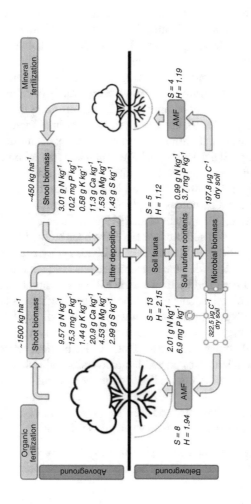

**Fig. 6.1** Long-term effects of mineral and organic fertilization on tropical soils and soil biota diversity. (Adapted from Souza and Santos (2018))

1. Continuous inputs of organic amendments and their residual effects on soil ecosystem
2. Balance between deposition, decomposition, and mineralization of soil organic matter
3. Habitat and energy provision for a wide range of soil organisms
4. Plant species fitness improvement

Many studies have showed that the use of manure and stubble retention can increase soil organic carbon to 40% after 3 years of its continuous use when compared to the conventional farming system (Li et al. 2018; Huang et al. 2019; Zhao et al. 2020). Other studies have reported similar phenomenon with total nitrogen. In organic sites, we found 21% more total nitrogen with stubble retention and organic inputs than in conventional farming system with mineral fertilization (Nunes et al. 2018; Adekiya et al. 2020). Both manure and mineral fertilizers have the advantage of supplying essential plant nutrients. However, they act in different ways into soil ecosystem. Usually, mineral fertilizer acts directly on plant nutrition with rapid response by increasing soil nutrient availability, but it disrupts in some cases the soil food web. On the other hand, organic fertilizers act indirectly on plant nutrition through nutrient cycling, but directly on soil biota function by alleviating soil pH and promoting habitat and resources supply for decomposers, litter transformers, and predators (Melo et al. 2019; Plaas et al. 2019; Mockeviciene et al. 2021).

The effects of land use on soil biodiversity have been studied in many countries around the world (Potapov et al. 2020; Yin et al. 2020; Turley et al. 2020; Wale and Yesuf 2021). Despite the considerable research that has been undertaken to measure these effects of land use on soil biota abundance, diversity, and functional aspects, predicting how these soil organisms respond to land use (e.g., monocropping, agro-forestry system, no-tillage, and organic tillage) remains difficult because of the wide variety of soil organism groups to identify specially the soil microbiota (Berkelmann et al. 2020; Méndez-Rojas et al. 2021). Further, we summarize the main effects of different land uses on soil biota diversity and richness (Table 6.2).

Basically, when comparing conventional farming systems with intensive land use and organic farming systems with traditional management, we can find two clear pathways related to the soil food web structure (Tian et al. 2017; Carrillo-Saucedo et al. 2018). The following pathways are as follows:

1. Bacterial-dominated soil food web: This pathway is characterized by high levels of icide-type products and fertilizer, high herbivory pressure, and reduced soil organic carbon inputs.
2. Fungal-dominated soil food web: This pathway is characterized by the continuous use of green manure, organic residues, and increased soil organic carbon inputs.

The soil microbial community into this context can be used as a bioindicator for land use intensity (Singh et al. 2020; Singh and Yadav 2021). Other studies have reported in organic farming systems a species-rich nematode and microarthropod community, especially fungal-feeders (e.g., fungivore nematodes, and springtails)

**Table 6.2**  Summary of main effects of different land uses on soil biota community structure

| Land use | Diversity (Shannon's index) | Dominance (Simpson's index) | Main findings |
|---|---|---|---|
| Conventional tillage | Low | High | By reducing soil organic matter content, conventional tillage reduces predator, litter transformers, and decomposer's diversity. This kind of system tends to promote ecosystem engineers' diversity (e.g., ants, larvae, and termites). We can find functional redundancy into conventional tillage |
| Organic tillage | High | High | By providing habitat and resource when using organic residues (incorporated or not into soil profile), this land use creates positive plant-soil feedback thus promoting the diversity of a wide range of soil organism' functional groups. Here, there is high dominance of ants and termites yet. After 3 years, organic tillage tends to restore the soil food web structure |
| No tillage | Low | Low | Without active primary producers, root exudation, and rhizodeposition, the soil food web tends to reduce its structure. First, we lost diversity, and last we lost dominance. It is also considered the first stage of degradation if any soil management practice has been applied |
| Agroforestry system | High | Rarely high | It tries to copy or simulate the natural ecosystem conditions. Generally, we need at least 5 years to start showing positive effect of an agroforestry system on soil food web. This system is characterized by high predators' presence and low herbivores abundance |
| Natural ecosystem (forest) | Low | High | Tropical forests tend to show high plant diversity, but for soil organisms, we just have few groups of soil organisms acting into the soil food web. These specific groups tend to present high abundance (arachnids, beetles, termites) and to provide important soil functions |
| Natural ecosystem (Savannah) | High | Low | Tropical Savannahs tend to show lower plant biomass when compared to tropical forests. It is related to its own plant traits. Here, we found high diversity of soil organisms without a specific dominance |
| Desert | Low | Low | The same process is observed in no tillage areas. However, the climatic traits potentialize the low diversity and dominance. There is high resilience |

Adapted from Ma et al. (2018), Xiao et al. (2019), Nanganoa et al. (2019), Aldebron et al. (2020), Boinot et al. (2020), Forstall-Sosa et al. (2020)

because of the fungal-dominated soil food web. Here, we must consider that in organic farming systems there is a high abundance of fungal species, thus promoting a positive plant-soil feedback into soil food web. On the other hand, conventional farming systems create a negative plant-soil feedback into soil food web, thus reducing both density and richness of fungal-feeders and promoting in some cases in tropical ecosystem root-knot nematodes (e.g., *Meloidogyne*) through functional redundancy process where the fungal-feeders are replaced by other opportunist herbivores and bacterial-feeders (Lupatini et al. 2019; Schmitt et al. 2021).

## 6.3  Soil Organism's Function in Contaminated Areas

The human activities and some farming systems have generated a wide range of chemical products and waste (Sun et al. 2019; Hou et al. 2020; Kafaei et al. 2020). Sometimes, these waste types are used into agriculture systems as soil promoters and fertilizers (e.g., organic residues). However, in other cases, these residues can promote toxic effect (e.g., *icide*-type products, heavy metals, and biohazard products) on soil organisms (Müller 2018; Mulla et al. 2019). In such cases, we consider these compounds as pollutants (Wołejko et al. 2020; Li et al. 2021b; Yan et al. 2021). Consequently, both plant species and soil biota are constantly exposed to these compounds that are both legally and illegally released into the environment, or they are used at high concentrations into some soil management practices (Sant'Ana et al. 2019; Malaj et al. 2020), thus affecting plant fitness and soil biota function (Tripathi et al. 2020; Pelosi et al. 2021).

In soil ecosystem, these compounds tend to accumulate over time into soil profile through the clay and organic matter capacity to bind them (Zhang et al. 2020; Sarkar et al. 2020). This process affects directly and indirectly soil biota (Ye et al. 2017; Geissen et al. 2021). As already described, soil organisms promote many functions in soil ecosystem (e.g., nutrient cycling, soil organic matter formation, soil structure, biological control). So, soil contamination can lead to soil food web imbalance which in a long timescale may compromise soil quality (Soliman et al. 2019; Tibbett et al. 2020). Commonly, the most affected soil organism group is the invertebrates (e.g., earthworms, springtails, oligochaetes, enchytraeids, and mites). They play an important role on soil structure, microregulation, and soil organic matter decomposition (Long et al. 2021; Tan et al. 2021), and many studies have reported negative effects of soil pollutants on their diversity, fitness, reproduction, and avoidance (Pereira et al. 2018). Other studies have showed changes on soil organisms' traits as affected by soil pollutants (DiBartolomeis et al. 2019; Sattler et al. 2020).

In tropical ecosystems, soil organism's function is essential and a useful tool as bioindicator of contaminated areas by pollutants (Pelosi and Römbke 2018; Christel et al. 2021). Some studies have used different approaches to show how soil organisms act in contaminated areas, as follows:

1. Germination test: This kind of test aims to access the viability of the soil seed bank or the individual fitness of a species seed type (e.g., leguminous, grasses, etc.). It is an important bioindicator that shows the natural regeneration and germination rate for a studied environment (Tran et al. 2018).

2. Tissue characterization: By sampling and analyzing the chemical composition of plant and animal tissues, scientific research may indicate if any pollutant is present in the soil food web. It is important to build management practices for a specific situation (e.g., Hg contamination) (Li et al. 2021b).
3. Reproduction test: It shows the organisms' reproduction in specific situations (contaminated vs. non-contaminated treatments). It is expected to find better results of reproduction in habitat that provide enough resources without disturbing the organism's ability to reproduce. Contaminated areas are commonly correlated with low reproduction rates. For this kind of test, we look to use mites, springtails, and earthworms because they are very sensitive to soil contamination (Princz et al. 2018).
4. Avoidance test: This test aims to show how is the affinity of any organisms to a specific soil condition (e.g., acid soil, low fertility, pollutant presence), but to study levels of soil contamination, we commonly use earthworms. This soil organisms' group is very sensitive to pollutant presence, and they always avoid contaminated areas. It is a quick test without any use of high technology (Ge et al. 2018) that provides good results combined with high accuracy.
5. Fitness test: This test combines a multiple approach into field studies. Here the researcher must test soil organisms' abundance, diversity, biomass, morphological, physiological, and anatomical traits. This is widely used worldwide for soil ecologists (Kristiansen et al. 2021), but its accuracy is sample and site dependent. Also, it could vary by seasonality, land use, and vegetation type.

## 6.4   Conclusion

At the tropics, land use and soil contamination are strongly linked to each other. The first one determines the balance of organic matter input, soil function, and soil organisms' diversity, while the second one could be a reaction of the land use that might affect soil organisms' function, reproduction, and behavior. Land use is also directly connected to net primary production and thus may influence soil biological properties. Thus, we must consider that land use with high diversity of primary producers (Agroforestry system) may present similar patterns of natural ecosystems (Villanueva-López et al. 2019; Laurindo et al. 2021). In fact, agroforestry systems provide habitat for a wide variety of soil organisms (Barrios et al. 2018; Sauvadet et al. 2019; Staton et al. 2021). These soil organisms can be used as bioindicators to test if there is any contamination degree in a specific area or ecosystem, and germination test, tissue characterization, reproduction, avoidance, and fitness tests are the main approaches that a soil ecologist must use to characterize contaminated and non-contaminated ecosystems.

# References

Adekiya AO, Ejue WS, Olayanju A, Dunsin O, Aboyeji CM, Aremu C, Adegbite K, Akinpelu O (2020) Different organic manure sources and NPK fertilizer on soil chemical properties, growth, yield and quality of okra. Sci Rep 10(1):1–9. https://doi.org/10.1038/s41598-020-73291-x

Aldebron C, Jones MS, Snyder WE, Blubaugh CK (2020) Soil organic matter links organic farming to enhanced predator evenness. Biol Control 146:104278. https://doi.org/10.1016/j.biocontrol.2020.104278

Alewell C, Ringeval B, Ballabio C, Robinson DA, Panagos P, Borrelli P (2020) Global phosphorus shortage will be aggravated by soil erosion. Nat Commun 11(1):1–12. https://doi.org/10.1038/s41467-020-18326-7

Alvarenga IFS, Dos Santos FE, Silveira GL, Andrade-Vieira LF, Martins GC, Guilherme LRG (2020) Investigating arsenic toxicity in tropical soils: a cell cycle and DNA fragmentation approach. Sci Total Environ 698:134272. https://doi.org/10.1016/j.scitotenv.2019.134272

Barbosa LS, Souza TAF, Lucena EO, da Silva LJR, Laurindo LK, Nascimento GS, Santos D (2021) Arbuscular mycorrhizal fungi diversity and transpiratory rate in long-term field cover crop systems from tropical ecosystem, northeastern Brazil. Symbiosis. https://doi.org/10.1007/s13199-021-00805-0

Barrios E, Valencia V, Jonsson M, Brauman A, Hairiah K, Mortimer PE, Okubo S (2018) Contribution of trees to the conservation of biodiversity and ecosystem services in agricultural landscapes. Int J Biodiver Sci Ecosyst Serv Manage 14(1):1–16. https://doi.org/10.1080/21513732.2017.1399167

Berkelmann D, Schneider D, Meryandini A, Daniel R (2020) Unravelling the effects of tropical land use conversion on the soil microbiome. Environ Microb 15(1):1–18. https://doi.org/10.1186/s40793-020-0353-3

Boinot S, Mézière D, Poulmarc'h J, Saintilan A, Lauri PE, Sarthou JP (2020) Promoting generalist predators of crop pests in alley cropping agroforestry fields: farming system matters. Ecol Eng 158:106041. https://doi.org/10.1016/j.ecoleng.2020.106041

Brunetti G, Traversa A, De Mastro F, Cocozza C (2019) Short term effects of synergistic inorganic and organic fertilization on soil properties and yield and quality of plum tomato. Sci Hortic 252:342–347. https://doi.org/10.1016/j.scienta.2019.04.002

Carrillo-Saucedo SM, Gavito ME, Siddique I (2018) Arbuscular mycorrhizal fungal spore communities of a tropical dry forest ecosystem show resilience to land-use change. Fungal Ecol 32:29–39. https://doi.org/10.1016/j.funeco.2017.11.006

Christel A, Maron PA, Ranjard L (2021) Impact of farming systems on soil ecological quality: a meta-analysis. Environ Chem Lett 19:4603–4625. https://doi.org/10.1007/s10311-021-01302-y

Daam MA, Chelinho S, Niemeyer JC, Owojori OJ, de Silva PMC, Sousa JP, van Gestel CAM, Römbke J (2019) Environmental risk assessment of pesticides in tropical terrestrial ecosystems: test procedures, current status and future perspectives. Ecotoxicol Environ Saf 181:534–547. https://doi.org/10.1016/j.ecoenv.2019.06.038

DalCorso G, Fasani E, Manara A, Visioli G, Furini A (2019) Heavy metal pollutions: state of the art and innovation in phytoremediation. Int J Mol Sci 20(14):3412. https://doi.org/10.3390/ijms20143412

Dangi S, Gao S, Duan Y, Wang D (2020) Soil microbial community structure affected by biochar and fertilizer sources. Appl Soil Ecol 150:103452. https://doi.org/10.1016/j.apsoil.2019.103452

DiBartolomeis M, Kegley S, Mineau P, Radford R, Klein K (2019) An assessment of acute insecticide toxicity loading (AITL) of chemical pesticides used on agricultural land in the United States. PLoS One 14(8):e0220029. https://doi.org/10.1371/journal.pone.0220029

Feng G, Xie T, Wang X, Bai J, Tang L, Zhao H, Wei W, Wang M, Zhao Y (2018) Metagenomic analysis of microbial community and function involved in cd-contaminated soil. BMC Microbiol 18:11. https://doi.org/10.1186/s12866-018-1152-5

Feng X, Wang Q, Sun Y, Zhang S, Wang F (2022) Microplastics change soil properties, heavy metal availability and bacterial community in a Pb-Zn-contaminated soil. J Hazard Mater 424:127364. https://doi.org/10.1016/j.jhazmat.2021.127364

Figlioli F, Sorrentino MC, Memoli V, Arena C, Maisto G, Giordano S, Capozzi F, Spagnuolo V (2019) Overall plant responses to Cd and Pb metal stress in maize: growth pattern, ultra-structure, and photosynthetic activity. Environ Sci Pollut Res 26(2):1781–1790. https://doi.org/10.1007/s11356-018-3743-y

Forstall-Sosa KS, Souza TAF, Lucena EO, da Silva SAI, Ferreira JTA, Silva TN, Santos D, Niemeyer JC (2020) Soil macroarthropod community and soil biological quality index in a green manure farming system of the Brazilian semi-arid. Biologia. https://doi.org/10.2478/s11756-020-00602-y

Franco AL, Cherubin MR, Cerri CE, Six J, Wall DH, Cerri CC (2020) Linking soil engineers, structural stability, and organic matter allocation to unravel soil carbon responses to land-use change. Soil Biol Biochem 150:107998. https://doi.org/10.1016/j.soilbio.2020.107998

Ge J, Xiao Y, Chai Y, Yan H, Wu R, Xin X, Wang D, Yu X (2018) Sub-lethal effects of six neonic-otinoids on avoidance behavior and reproduction of earthworms (Eisenia fetida). Ecotoxicol Environ Saf 162:423–429. https://doi.org/10.1016/j.ecoenv.2018.06.064

Geisen S, Wall DH, van der Puten WH (2019) Challenges and opportunities for soil biodiversity in the Anthropocene. Curr Biol 29(19):R1036–R1044. https://doi.org/10.1016/j.cub.2019.08.007

Geissen V, Silva V, Lwanga EH, Beriot N, Oostindie K, Bin Z, Pyne E, Busink S, Zomer P, Mol H, Ritsema CJ (2021) Cocktails of pesticide residues in conventional and organic farming systems in Europe–Legacy of the past and turning point for the future. Environ Pollut 278: 116827. https://doi.org/10.1016/j.envpol.2021.116827

Gomes AR, Justino C, Rocha-Santos T, Freitas AC, Duarte AC, Pereira R (2017) Review of the ecotoxicological effects of emerging contaminants to soil biota. J Environ Sci Health 52(10):992–1007. https://doi.org/10.1080/10934529.2017.1328946

Harkes P, Suleiman AKA, van den Elsen SJJ, de Haan JJ, Holterman M, Kuramae EE, Helder J (2019) Conventional and organic soil management as divergent drivers of resident and active fractions of major soil food web constituents. Sci Rep 9:13521. https://doi.org/10.1038/s41598-019-49854-y

Hossain A, Krupnik TJ, Timsina J, Mahboob MG, Chaki AK, Farooq M, Bhatt R, Fahad S, Hasanuzzaman M (2020) Agricultural land degradation: processes and problems undermining future food security. In: Fahad S, Hasanuzzaman M, Alam M, Ullah H, Saeed M, Ali Khan I, Adnan M (eds) Environment, climate, plant and vegetation growth. Springer, Cham. https://doi.org/10.1007/978-3-030-49732-3_2

Hou D, O'Connor D, Igalavithana AD, Alessi DS, Luo J, Tsang DCW, Sparks DL, Yamauchi Y, Rinklebe J, Ok YS (2020) Metal contamination and bioremediation of agricultural soils for food safety and sustainability. Nat Rev Earth Environ. https://doi.org/10.1038/s43017-020-0061-y

Huang X, Jia Z, Guo J, Li T, Sun D, Meng H, Yu G, He X, Ran W, Zhang S, Hong J, Shen Q (2019) Ten-year long-term organic fertilization enhances carbon sequestration and calcium-mediated stabilization of aggregate-associated organic carbon in a reclaimed Cambisol. Geoderma 355:113880. https://doi.org/10.1016/j.geoderma.2019.113880

Huang R, Gao X, Wang F, Xu G, Long Y, Wang C, Wang Z, Gao M (2020) Effects of biochar incorporation and fertilizations on nitrogen and phosphorus losses through surface and sub-surface flows in a sloping farmland of Entisol. Agric Ecosyst Environ 300:106988. https://doi.org/10.1016/j.agee.2020.106988

Ikoyi I, Fowler A, Storey S, Doyle E, Schmalenberger A (2020) Sulfate fertilization supports growth of ryegrass in soil columns but changes microbial community structures and reduces abundances of nematodes and arbuscular mycorrhiza. Sci Total Environ 704:135315. https://doi.org/10.1016/j.scitotenv.2019.135315

Kafaei R, Arfaeinia H, Savari A, Mahmoodi M, Rezaei M, Rayani M, Sorial GA, Fattahi N, Ramavandi B (2020) Organochlorine pesticides contamination in agricultural soils of southern Iran. Chemosphere 240:124983. https://doi.org/10.1016/j.chemosphere.2019.124983

Kristiansen SM, Borgå K, Rundberget JT, Leinaas HP (2021) Effects on life-history traits of *Hypogastrura viatica* (Collembola) exposed to imidacloprid through soil or diet. Environ Toxicol Chem 40(11):3111–3122. https://doi.org/10.1002/etc.5187

Laurindo LK, Souza TAF, da Silva LJR, Casal TB, Pires KDJC, Kormann S, Schmitt DE, Siminski A (2021) Arbuscular mycorrhizal fungal community assembly in agroforestry systems from the Southern Brazil. Biologia 76(4):1099–1107. https://doi.org/10.1007/s11756-021-00700-5

Lecerf A, Cébron A, Gilbert F, Danger M, Roussel H, Maunoury-Danger F (2021) Using plant litter decomposition as an indicator of ecosystem response to soil contamination. Ecol Indic 125:107554. https://doi.org/10.1016/j.ecolind.2021.107554

Li J, Wen Y, Li X, Li Y, Yang X, Lin Z, Song Z, Cooper JM, Zhao B (2018) Soil labile organic carbon fractions and soil organic carbon stocks as affected by long-term organic and mineral fertilization regimes in the North China Plain. Soil Tillage Res 175:281–290. https://doi.org/10.1016/j.still.2017.08.008

Li Y, Chen Y, Li J, Sun Q, Li J, Xu J, Liu B, Lang Q, Qiao Y (2021a) Organic management practices enhance soil food web biomass and complexity under greenhouse conditions. Appl Soil Ecol 167:104010. https://doi.org/10.1016/j.apsoil.2021.104010

Li C, Xu Z, Luo K, Chen Z, Xu X, Xu C, Qiu G (2021b) Biomagnification and trophic transfer of total mercury and methylmercury in a sub-tropical montane forest food web, Southwest China. Chemosphere 277:130371. https://doi.org/10.1016/j.chemosphere.2021.130371

Lin Y, Ye G, Kuzyakov Y, Liu D, Fan J, Ding W (2019) Long-term manure application increases soil organic matter and aggregation, and alters microbial community structure and keystone taxa. Soil Biol Biochem 134:187–196. https://doi.org/10.1016/j.soilbio.2019.03.030

Liu S, Zamanian K, Schleuss PM, Zarebanadkouki M, Kuzyakov Y (2018) Degradation of Tibetan grasslands: consequences for carbon and nutrient cycles. Agric Ecosyst Environ 252:93–104. https://doi.org/10.1016/j.agee.2017.10.011

Long J, Zhang M, Li J, Liao H, Wang X (2021) Soil macro- and mesofauna-mediated litter decomposition in a subtropical karst forest. Biotropica. https://doi.org/10.1111/btp.12980

Lori M, Symnaczik S, Mäder P, De Deyn G, Gattinger A (2017) Organic farming enhances soil microbial abundance and activity—a meta-analysis and meta-regression. PLoS One 12(7):e0180442. https://doi.org/10.1371/journal.pone.0180442

Lupatini M, Korthals GW, Roesch LF, Kuramae EE (2019) Long-term farming systems modulate multi-trophic responses. Sci Total Environ 646:480–490. https://doi.org/10.1016/j.scitotenv.2018.07.323

Lykogianni M, Bempelou E, Karamaouna F, Aliferis KA (2021) Do pesticides promote or hinder sustainability in agriculture? The challenge of sustainable use of pesticides in modern agriculture. Sci Total Environ 795:148625. https://doi.org/10.1016/j.scitotenv.2021.148625

Ma Q, Yu H, Liu X, Xu Z, Zhou G, Shi Y (2018) Climatic warming shifts the soil nematode community in a desert steppe. Clim Chang 150:243–258. https://doi.org/10.1007/s10584-018-2277-0

Malaj E, Liber K, Morrissey CA (2020) Spatial distribution of agricultural pesticide use and predicted wetland exposure in the Canadian Prairie Pothole Region. Sci Total Environ 718:134765. https://doi.org/10.1016/j.scitotenv.2019.134765

Martínez-García LB, Korthals G, Brussaard L, Jørgensen HB, De Deyn GB (2018) Organic management and cover crop species steer soil microbial community structure and functionality along with soil organic matter properties. Agric Ecosyst Environ 263:7–17. https://doi.org/10.1016/j.agee.2018.04.018

Martins JJF, Soares AM, Azeiteiro UM, Correia MLT (2019) Anthropic action effects caused by soybean farmers in a watershed of Tocantins—Brazil and its connections with climate change. In: Castro P, Azul A, Leal Filho W, Azeiteiro U (eds) Climate change-resilient agriculture and agroforestry. Climate change management. Springer, Cham. https://doi.org/10.1007/978-3-319-75004-0_15

Medorio-García HP, Alarcón E, Flores-Esteves N, Montaño NM, Perroni Y (2020) Soil carbon, nitrogen and phosphorus dynamics in sugarcane plantations converted from tropical dry forest. Appl Soil Ecol 154:103600. https://doi.org/10.1016/j.apsoil.2020.103600

Melo LN, Souza TAF, Santos D (2019) Cover crop farming systemaffect macroarthropods community diversity of Caatinga Brazil. Biologia. https://doi.org/10.2478/s11756-019-00272-5

Méndez-Rojas DM, Cultid-Medina C, Escobar F (2021) Influence of land use change on rove beetle diversity: a systematic review and global meta-analysis of a mega-diverse insect group. Ecol Indic 122:107239. https://doi.org/10.1016/j.ecolind.2020.107239

Menta C, Remelli S (2020) Soil health and arthropods: from complex system to worthwhile investigation. Insects 11:54. https://doi.org/10.3390/insects11010054

Menta C, Conti FD, Lozano Fondón C, Staffilani F, Remelli S (2020) Soil arthropod responses in agroecosystem: implications of different management and cropping systems. Agronomy 10(7):982. https://doi.org/10.3390/agronomy10070982

Mockeviciene I, Repsiene R, Amaleviciute-Volunge K, Karcauskiene D, Slepetiene A, Lepane V (2021) Effect of long-term application of organic fertilizers on improving organic matter quality in acid soil. Arch Agron Soil Sci. https://doi.org/10.1080/03650340.2021.1875130

Morgado RG, Loureiro S, González-Alcaraz MN (2018) Changes in soil ecosystem structure and functions due to soil contamination. In: Duarte AC, Cachada A, Rocha-Santos T (eds) Soil pollution. Elsevier, pp 59–87. https://doi.org/10.1016/B978-0-12-849873-6.00003-0

Muchane MN, Sileshi GW, Gripenberg S, Jonsson M, Pumariño L, Barrios E (2020) Agroforestry boosts soil health in the humid and sub-humid tropics: a meta-analysis. Agric Ecosyst Environ 295:106899. https://doi.org/10.1016/j.agee.2020.106899

Mulla SI, Ameen F, Talwar MP, Eqani SAMAS, Bharagava RN, Saxena G, Tallur PN, Ninnekar HZ (2019) Organophosphate pesticides: impact on environment, toxicity, and their degradation. In: Saxena G, Bharagava R (eds) Bioremediation of industrial waste for environmental safety. Springer, Singapore. https://doi.org/10.1007/978-981-13-1891-7_13

Müller C (2018) Impacts of sublethal insecticide exposure on insects—facts and knowledge gaps. Basic Appl Ecol 30:1–10. https://doi.org/10.1016/j.baae.2018.05.001

Nanganoa LT, Okolle JN, Missi V, Tueche JR, Levai LD, Njukeng JN (2019) Impact of different land-use systems on soil physicochemical properties and macrofauna abundance in the humid tropics of Cameroon. Appl Environ Soil Sci. https://doi.org/10.1155/2019/5701278

Nascimento GS, Souza TAF, Silva LJR, Santos D (2021) Soil physico-chemical properties, biomass production, and root density in a green manure farming system from tropical ecosystem, North-eastern Brazil. J Soils Sediments. https://doi.org/10.1007/s11368-021-02924-z

Ning Q, Chen L, Jia Z, Zhang C, Ma D, Li F, Zhang J, Li D, Han X, Cai Z, Huang S, Liu W, Zhu B, Li Y (2020) Multiple long-term observations reveal a strategy for soil pH-dependent fertilization and fungal communities in support of agricultural production. Agric Ecosyst Environ 293:106837. https://doi.org/10.1016/j.agee.2020.106837

Nunes MR, van Es HM, Schindelbeck R, Ristow AJ, Ryan M (2018) No-till and cropping system diversification improve soil health and crop yield. Geoderma 328:30–43. https://doi.org/10.1016/j.geoderma.2018.04.031

Olorunfemi IE, Komolafe AA, Fasinmirin JT, Olufayo AA (2019) Biomass carbon stocks of different land use management in the forest vegetative zone of Nigeria. Acta Oecol 95:45–56. https://doi.org/10.1016/j.actao.2019.01.004

Oshunsanya SO, Yu H, Li Y (2018) Soil loss due to root crop harvesting increases with tillage operations. Soil Tillage Res 181:93–101. https://doi.org/10.1016/j.still.2018.04.003

Oumenskou H, El Baghdadi M, Barakat A, Aquit M, Ennaji W, Karroum LA, Aadraoui M (2018) Assessment of the heavy metal contamination using GIS-based approach and pollution indices in agricultural soils from Beni Amir irrigated perimeter, Tadla plain, Morocco. Arab J Geosci 11:692. https://doi.org/10.1007/s12517-018-4021-5

Ozlu E, Kumar S (2018) Response of soil organic carbon, pH, electrical conductivity, and water stable aggregates to long-term annual manure and inorganic fertilizer. Soil Sci Soc Am J 82(5):1243–1251. https://doi.org/10.2136/sssaj2018.02.0082

Pelosi C, Römbke J (2018) Enchytraeids as bioindicators of land use and management. Appl Soil Ecol 123:775–779. https://doi.org/10.1016/j.apsoil.2017.05.014

Pelosi C, Bertrand C, Daniele G, Coeurdassier M, Benoit P, Nélieu S, Lafay F, Bretagnolle V, Gaba S, Vulliet E, Fritsch C (2021) Residues of currently used pesticides in soils and earthworms: a silent threat? Agric Ecosyst Environ 305:107167. https://doi.org/10.1016/j.agee.2020.107167

Pereira R, Cachada A, Sousa JP, Niemeyer J, Markwiese J, Andersen CP (2018) Ecotoxicological effects and risk assessment of pollutants. In: Duarte AC, Cachada A, Rocha-Santos T (Eds.) Soil pollution from monitoring to remediation. Elsevier pp. 191–216. https://doi.org/10.1016/B978-0-12-849873-6.00008-X

Plaas E, Meyer-Wolfarth F, Banse M, Bengtsson J, Bergmann H, Faber J, Potthoff M, Runge T, Schrader S, Taylor A (2019) Towards valuation of biodiversity in agricultural soils: a case for earthworms. Ecol Econ 159:291–300. https://doi.org/10.1016/j.ecolecon.2019.02.003

Potapov AM, Dupérré N, Jochum M, Dreczko K, Klarner B, Barnes AD, Krashevska V, Rembold K, Kreft H, Brose U, Widyastuti R, Harms D, Scheu S (2020) Functional losses in ground spider communities due to habitat structure degradation under tropical land-use change. Ecology 101(3):e02957. https://doi.org/10.1002/ecy.2957

Princz J, Jatar M, Lemieux H, Scroggins R (2018) Perfluorooctane sulfonate in surface soils: effects on reproduction in the collembolan, *Folsomia candida*, and the oribatid mite, *Oppia nitens*. Chemosphere 208:757–763. https://doi.org/10.1016/j.chemosphere.2018.06.020

Puissant J, Villenave C, Chauvin C, Plassard C, Blanchart E, Trap J (2021) Quantification of the global impact of agricultural practices on soil nematodes: a meta-analysis. Soil Biol Biochem 161:108383. https://doi.org/10.1016/j.soilbio.2021.108383

Purwanto BH, Alam S (2020) Impact of intensive agricultural management on carbon and nitrogen dynamics in the humid tropics. Soil Sci Plant Nutr 66(1):50–59. https://doi.org/10.1080/00380768.2019.1705182

Ramirez-Morales D, Perez-Villanueva ME, Chin-Pampillo JS, Aguilar-Mora P, Arias-Mora V, Masis-Mora M (2021) Pesticide occurrence and water quality assessment from an agriculturally influenced Latin-American tropical region. Chemosphere 262:127851. https://doi.org/10.1016/j.chemosphere.2020.127851

Raza S, Miao N, Wang P, Ju X, Chen Z, Zhou J, Kuzyakov Y (2020) Dramatic loss of inorganic carbon by nitrogen-induced soil acidification in Chinese croplands. Glob Chang Biol 26(6):3738–3751. https://doi.org/10.1111/gcb.15101

Renaud M, Akeju T, Natal-da-Luz T, Leston S, Rosa J, Ramos F, Sousa JP, Azevedo-Pereira HMVS (2018) Effects of the neonicotinoids acetamiprid and thiacloprid in their commercial formulations on soil fauna. Chemosphere 194:85–93. https://doi.org/10.1016/j.chemosphere.2017.11.102

Sanaullah M, Usman M, Wakeel A, Cheema SA, Ashraf I, Farooq M (2020) Terrestrial ecosystem functioning affected by agricultural management systems: a review. Soil Tillage Res 196:104464. https://doi.org/10.1016/j.still.2019.104464

Sánchez-Bayo F (2021) Indirect effect of pesticides on insects and other arthropods. Toxics 9(8):177. https://doi.org/10.3390/toxics9080177

Sant'Ana G, SHC A, Jardel OEE (2019) Apprehension of illegal pesticides, agricultural productivity and food poisoning on the Brazilian state of Mato Grosso do Sul. Revista de Ciencias Agrícolas 36:52–62. https://doi.org/10.22267/rcia.1936e.106

Sarkar B, Mukhopadhyay R, Mandal A, Mandal S, Vithanage M, Biswas JK (2020) Sorption and desorption of agro-pesticides in soils. In: Prasad MNV (ed) Agrochemicals detection. Treat Remed, Elsevier, pp 189–205. https://doi.org/10.1016/B978-0-08-103017-2.00008-8

Sattler C, Gianuca AT, Schweiger O, Franzén M, Settele J (2020) Pesticides and land cover heterogeneity affect functional group and taxonomic diversity of arthropods in rice agroecosystems. Agric Ecosyst Environ 297:106927. https://doi.org/10.1016/j.agee.2020.106927

Sauvadet M, Van den Meersche K, Allinne C, Gay F, de Melo Virginio Filho E, Chauvat M, Becqueer T, Tixier P, Harmand JM (2019) Shade trees have higher impact on soil nutrient availability and food web in organic than conventional coffee agroforestry. Sci Total Environ 649:1065–1074. https://doi.org/10.1016/j.scitotenv.2018.08.291

Schmitt J, Portela VO, Santana NA, Freiberg JA, Bellé C, Pacheco D, Antoniolli ZI, Anghioni I, Cares JE, Araújo Filho JV, Jacques RJS (2021) Effect of grazing intensity on plant-parasitic nematodes in an integrated crop–livestock system with low plant diversity. Appl Soil Ecol 163:103908. https://doi.org/10.1016/j.apsoil.2021.103908

Schrama M, de Haan JJ, Kroonen M, Vestegen H, van der Putten WH (2018) Crop yield gap and stability in organic and conventional farming systems. Agri Ecosyst Environ 256:123–130. https://doi.org/10.1016/j.agee.2017.12.023

Siebert J, Sünnemann M, Auge H, Berger S, Cesarz S, Ciobanu M, Guerrero-Ramírez NR, Eisenhauer N (2019) The effects of drought and nutrient addition on soil organisms vary across taxonomic groups, but are constant across seasons. Sci Rep 9(1):1–12. https://doi.org/10.1038/s41598-018-36777-3

Siebert J, Ciobanu M, Schädler M, Eisenhauer N (2020) Climate change and land use induce functional shifts in soil nematode communities. Oecologia 192:281–294. https://doi.org/10.1007/s00442-019-04560-4

Singh B (2018) Are nitrogen fertilizers deleterious to soil health? Agronomy 8(4):48. https://doi.org/10.3390/agronomy8040048

Singh S, Yadav H (2021) Influence of land use change on native microbial community and their response to the variations in micro environment. In: Singh C, Tiwari S, Singh AK (eds) Singh JS. Elsevier, Microbes in land use change management, pp 325–340. https://doi.org/10.1016/B978-0-12-824448-7.00018-8

Singh AK, Jiang XJ, Yang B, Wu J, Rai A, Chen C, Ahirwal J, Wang P, Liu W, Singh N (2020) Biological indicators affected by land use change, soil resource availability and seasonality in dry tropics. Ecol Indic 115:106369. https://doi.org/10.1016/j.ecolind.2020.106369

Sofo A, Mininni AN, Ricciuti P (2020) Soil macrofauna: a key factor for increasing soil fertility and promoting sustainable soil use in fruit orchard agrosystems. Agronomy 10(4):456. https://doi.org/10.3390/agronomy10040456

Soliman M, El-Shazly M, Abd-El-Samie E, Fayed H (2019) Variations in heavy metal concentrations among trophic levels of the food webs in two agroecosystems. Afr Zool 54:21–30. https://doi.org/10.1080/15627020.2019.1583080

Souza TAF, de Andrade LA, Freitas H, Sandim AS (2018) Biological invasion influences the outcome of plant-soil feedback in the invasive plant species from the Brazilian semi-arid. Microb Ecol 76:102–112. https://doi.org/10.1007/s00248-017-0999-6

Staton T, Walters RJ, Smith J, Breeze TD, Girling RD (2021) Evaluating a trait-based approach to compare natural enemy and pest communities in agroforestry vs. arable systems. Ecol Appl 31(4):e02294. https://doi.org/10.1002/eap.2294

Sun Y, Li H, Guo G, Semple KT, Jones KC (2019) Soil contamination in China: current priorities, defining background levels and standards for heavy metals. J Environ Manag 251:109512. https://doi.org/10.1016/j.jenvman.2019.109512

Tan B, Yin R, Zhang J, Xu Z, Liu Y, He S, Zhang L, Li H, Wang L, Liu S, You C, Peng C (2021) Temperature and moisture modulate the contribution of soil fauna to litter decomposition via different pathways. Ecosystems 24:1142–1156. https://doi.org/10.1007/s10021-020-00573-w

Tang J, Zhang J, Ren L, Zhou Y, Gao J, Luo L, Yang Y, Peng Q, Huang H, Chen A (2019) Diagnosis of soil contamination using microbiological indices: a review on heavy metal pollution. J Environ Manag 242:121–130. https://doi.org/10.1016/j.jenvman.2019.04.061

Thangavel P, Sridevi G (2017) Soil security: a key role for sustainable food productivity. In: Dhanarajan A (ed) Sustainable agriculture towards food security. Springer, Singapore. https://doi.org/10.1007/978-981-10-6647-4_16

Thomaz E, Antoneli V (2021) Long-term soil quality decline due to the conventional tobacco tillage in southern Brazil. Arch Agron Soil Sci. https://doi.org/10.1080/03650340.2020.1852550

Tian Q, Taniguchi T, Shi WY, Li G, Yamanaka N, Du S (2017) Land-use types and soil chemical properties influence soil microbial communities in the semiarid Loess Plateau region in China. Sci Rep 7:45289. https://doi.org/10.1038/srep45289

Tibbett M, Fraser TD, Duddigan S (2020) Identifying potential threats to soil biodiversity. PeerJ 8:e9271. https://doi.org/10.7717/peerj.9271

Tran TH, Gati EM, Eshel A, Winters G (2018) Germination, physiological and biochemical responses of acacia seedlings (*Acacia raddiana* and *Acacia tortilis*) to petroleum contaminated soils. Environ Pollut 234:642–655. https://doi.org/10.1016/j.envpol.2017.11.067

Tripathi S, Srivastava P, Devi RS, Bhadouria R (2020) Influence of synthetic fertilizers and pesticides on soil health and soil microbiology. In: Prasad MNV (ed) Agrochemicals detection. Treatment and Remediation, Elsevier pp, pp 25–54. https://doi.org/10.1016/B978-0-08-103017-2.00002-7

Tsai W-T (2019) Trends in the use of glyphosate herbicide and its relevant regulations in Taiwan: a water contaminant of increasing concern. Toxics 7:4. https://doi.org/10.3390/toxics7010004

Turley NE, Bell-Dereske L, Evans SE, Brudvig LA (2020) Agricultural land-use history and restoration impact soil microbial biodiversity. J Appl Ecol 57(5):852–863. https://doi.org/10.1111/1365-2664.13591

Van der Bom F, Nunes I, Raymond NS, Hansen V, Bonnichsen L, Magid J, Nybroe O, Jensen LS (2018) Long-term fertilisation form, level and duration affect the diversity, structure and functioning of soil microbial communities in the field. Soil Biol Biochem 122:91–103. https://doi.org/10.1016/j.soilbio.2018.04.003

Villanueva-López G, Lara-Pérez LA, Oros-Ortega I, Ramirez-Barajas PJ, Casanova-Lugo F, Ramos-Reyes R, Aryal DR (2019) Diversity of soil macro-arthropods correlates to the richness of plant species in traditional agroforestry systems in the humid tropics of Mexico. Agric Ecosyst Environ 286:106658. https://doi.org/10.1016/j.agee.2019.106658

Wale M, Yesuf S (2021) Abundance and diversity of soil arthropods in disturbed and undisturbed ecosystem in Western Amhara, Ethiopia. Int J Trop Insect Sci. https://doi.org/10.1007/s42690-021-00600-w

Wang Y, Wang ZL, Zhang Q, Hu N, Li Z, Lou Y, Li Y, Xue D, Chen Y, Wu C, Zou CB, Kuzyakov Y (2018) Long-term effects of nitrogen fertilization on aggregation and localization of carbon, nitrogen and microbial activities in soil. Sci Total Environ 624:1131–1139. https://doi.org/10.1016/j.scitotenv.2017.12.113

Wołejko E, Jabłońska-Trypuć A, Wydro U, Butarewicz A, Łozowicka B (2020) Soil biological activity as an indicator of soil pollution with pesticides–a review. Appl Soil Ecol 147:103356. https://doi.org/10.1016/j.apsoil.2019.09.006

Xiao D, Xiao S, Ye Y, Zhang W, He X, Wang K (2019) Microbial biomass, metabolic functional diversity, and activity are affected differently by tillage disturbance and maize planting in a typical karst calcareous soil. J Soils Sediments 19(2):809–821. https://doi.org/10.1007/s11368-018-2101-5

Yadav RP, Gupta B, Bhutia PL, Bisht JK, Pattanayak A (2019) Biomass and carbon budgeting of land use types along elevation gradient in Central Himalayas. J Clean Prod 211:1284–1298. https://doi.org/10.1016/j.jclepro.2018.11.278

Yan X, Wang J, Zhu L, Wang J, Li S, Kim YM (2021) Oxidative stress, growth inhibition, and DNA damage in earthworms induced by the combined pollution of typical neonicotinoid insecticides and heavy metals. Sci Total Environ 754:141873. https://doi.org/10.1016/j.scitotenv.2020.141873

Yang T, Siddique KH, Liu K (2020) Cropping systems in agriculture and their impact on soil health-a review. Glob Ecol Conserv 23:e01118. https://doi.org/10.1016/j.gecco.2020.e01118

Yang B, Banerjee S, Herzog C, Ramírez AC, Dahlin P, van der Heijden MG (2021) Impact of land use type and organic farming on the abundance, diversity, community composition and functional properties of soil nematode communities in vegetable farming. Agric Ecosyst Environ 318:107488. https://doi.org/10.1016/j.agee.2021.107488

Ye S, Zeng G, Wu H, Zhang C, Liang J, Dai J, Liu Z, Xiong W, Wan J, Xu P, Cheng M (2017) Co-occurrence and interactions of pollutants, and their impacts on soil remediation—a review. Crit Rev Environ Sci Technol 47(16):1528–1553. https://doi.org/10.1080/10643389.2017.1386951

Yin R, Kardol P, Thakur MP, Gruss I, Wu GL, Eisenhauer N, Schädler M (2020) Soil functional biodiversity and biological quality under threat: intensive land use outweighs climate change. Soil Biol Biochem 147:107847. https://doi.org/10.1016/j.soilbio.2020.107847

Zhang H, Yuan X, Xiong T, Wang H, Jiang L (2020) Bioremediation of co-contaminated soil with heavy metals and pesticides: influence factors, mechanisms and evaluation methods. Chem Eng J 398:125657. https://doi.org/10.1016/j.cej.2020.125657

Zhao ZB, He JZ, Geisen S, Han LL, Wang JT, Shen JP, Wei W, Fang T, Li P, Zhang LM (2019) Protist communities are more sensitive to nitrogen fertilization than other microorganisms in diverse agricultural soils. Microbiome 7(1):1–16. https://doi.org/10.1186/s40168-019-0647-0

Zhao Z, Zhang C, Li F, Gao S, Zhang J (2020) Effect of compost and inorganic fertilizer on organic carbon and activities of carbon cycle enzymes in aggregates of an intensively cultivated vertisol. PLoS One 15(3):e0229644. https://doi.org/10.1371/journal.pone.0229644

# Chapter 7
# Natural Ecosystems and Biological Invasion

**Abstract** This chapter explores how natural and invaded ecosystem provide habitat and energy supply for the entire soil food web, how biological invasion changes habitat of the soil organisms, and two study cases considering invasive plant species (*Cryptostegia madagascariensis* and *Prosopis juliflora*) from tropical zones. Natural ecosystem is defined as a community of biotic and abiotic entities that naturally occurs in a specific range, while biological invasion defines the spread and dominance of any organisms in a new range. These two concepts are strongly linked to each other in moist and dry tropical ecosystems, and in some cases, they create a war condition (by antagonism) that affects the entire soil food web. Natural ecosystem can provide a wide range of physical, chemical, and biological processes that promotes the entire soil food web, while invasive organisms just change the habitat for their own benefit.

**Keywords** Natural ecosystems · Native living organisms · Invasive organisms · Plant-soil feedback · Soil food web

**Questions Covered in the Chapter**
1. What is a natural ecosystem?
2. How can we define biological invasion?
3. How must soil food web works into moist and dry tropical ecosystem under natural and invaded conditions?
4. What does "habitat simplification" means and when does it occurs?
5. What are the main impacts of *Prosopis juliflora* and *Cryptostegia madagascariensis* invasion in the Caatinga ecoregion?

## 7.1 Introduction

When we started scientific studies considering soil ecology, it is important to characterize the studied environment (date, history, main habitats, their ecological traits), if it is a natural ecosystem, or if it is under any degree of biological invasion

© The Author(s), under exclusive license to Springer Nature Switzerland AG 2022    99
T. Souza, *Soil Biology in Tropical Ecosystems*, https://doi.org/10.1007/978-3-031-00949-5_7

(e.g., when an invasive exotic species creates monodominance and negatively affects native community composition) (Adigbli et al. 2019; Rai and Singh 2020; Bempah et al. 2021). It will clarify many aspects during our tests or analyses, and it will support our actions through the whole study (Thapa et al. 2018; Yletyinen et al. 2021). In this chapter, our aim was to show how natural and invaded ecosystem provide habitat and energy supply for the entire soil food web, how biological invasion changes habitat of the soil organisms, and two study cases considering invasive plant species (*Pinus* and *Acacia*) from tropical zones.

First, we must ask what is a natural ecosystem? It is defined as a complex community of biotic and abiotic entities that naturally occurs in a specific range (e.g., forests, savannahs, and deserts). Every entity interacts together through habitat provision and nests creation (physical process) and food and energy supply (chemical processes) and by establishing a predator-prey relationship (biological process and one of the most definitions of niche) (Mora 2018; Tsujimoto et al. 2018; Baruch et al. 2021) that creates positive plant-soil feedback overtime (De Long et al. 2019; Kareiva and Carranza 2018). In natural ecosystems, there is no or not significant human activity, and natural barriers of dispersion create a range where endemic and native organisms build their communities and establish evolutionary relationships together, thus creating adaptative, reproductive, and symbiotic traits (Maxwell et al. 2020; Wang et al. 2020).

However, what does occur if the human activity (e.g., agriculture, aquaculture, biocontrol, ornamentation, research, and transport) starts being significative? So, for this question the answer is: Biological invasions begin! Biological invasion is the result of the intentional and/or accidental spread of living organisms (Gao et al. 2018; Shackleton et al. 2019; Bertelsmeier 2021). It starts as a series of stages (transport, establishment, dispersion, and human perception) that living organisms must face to overcome the natural barriers of dispersion. The human activities can help (by improving speed and spatial scale of movement) this spread, thus breaking the natural barriers of dispersion (Hulme 2020; Pyšek et al. 2020). These kinds of living organisms that are introduced in new ranges can be classified accordingly to their behavior and spread characteristics:

1. Introduced, not native, or exotic species: It spreads outside its natural range which strongly depends to the human activity. It does not show any negative effects on natural ecosystem and just colonizes specific areas (e.g., agricultural fields, gardens, and city zoos).
2. Invasive species: It is dispersed widely, colonizing and invading the natural ecosystem, creating monodominance, disturbing the native community composition, and changing ecosystem services to its own benefit.

There are some studies showing the negative effects of biological invasion around the world (Cazetta and Zenni 2020; Jo et al. 2020; Lucena et al. 2021; Novoa et al. 2021). This process is considered one of the most serious threats for global biodiversity, and in tropical ecosystems, the invasive species can act as "evil" ecosystem engineers, because it can change both biotic and abiotic traits that can have strong impacts on the community structure (Ferlian et al. 2018; Frelich et al. 2019). The main impacts are as follows:

1. Direct impacts: The invader can affect the entire ecosystem through deep changes produced in the soil food web by changing parasitism, symbiosis, predation, and competition relationships (Zhang et al. 2018; Abgrall et al. 2019; Siddiqui et al. 2021). Other examples include changes on soil properties and nutrient cycling (Stefanowicz et al. 2018; Sun et al. 2019).
2. Indirect impacts: The invader can affect the interaction between the native living organisms on the ecosystem, sharing hosts (e.g., the case of *Funneliformis mosseae* and arbuscular mycorrhizal fungi for *Prosopis juliflora*, an invasive plant species from Brazilian tropical dry forest) and parasites (Sousa et al. 2017; Eshete et al. 2020).

The ecosystem function of invaded environments is very compromised (Linders et al. 2019; Seeney et al. 2020; Weidlich et al. 2020) because the invasive species starts building a new food web that the only aim it is to benefit itself. While natural ecosystem provides habitat and food resources for a wide range of living organisms that coevolved together, the invaded will change the entire ecosystem to provide favorable conditions to its prole (Stefanowicz et al. 2019; Milanović et al. 2020; Rodríguez-Caballero et al. 2020). So, we must consider several problems related to the biological invasion (Table 7.1) that can become extremely dangerous for our survival, for soil ecology, and for the natural ecosystem (Heringer et al. 2019; Wang et al. 2021a).

In fact, both natural and invaded ecosystem exert multiple effects on soil ecosystem, potentially affecting living organisms' interactions and global ecological processes (Dairel and Fidelis 2020; Torres et al. 2021; Řezáčová et al. 2021). They have a tremendous effect on soil biodiversity (Uboni et al. 2019; Renčo et al. 2021). It has been shown that the natural ecosystem substantially creates the most diverse organism community, whereas the invasive ecosystem changes soil communities

**Table 7.1** Main effects of long-term conventional farming system on ecosystem services at the tropics

| Main invader | Main effect | Soil biota compromised |
|---|---|---|
| Plant species | Less native plant species diversity (by killing native plant species through allelopathic processes) | Primary producers |
| | High litter deposition (creating physical barrier that compromises native soil seed bank) | Ecosystem engineers |
| | | Litter transformers |
| | High soil fertility (island of fertility hypothesis) | Prokaryotic transformers |
| | High growth and seed dispersion (no natural enemies) | Symbionts |
| Animal species | Disrupting the predator-prey relationship (by altering herbivory effects) | Herbivores |
| | | Predators |
| | Extinction events (by high predation pressure) | Decomposers |
| | High functional redundancy (by changing trophic levels) | Microregulators |

Adapted from Delon and Purves (2018), Zhang et al. (2018), Souza et al. (2019), Battini et al. (2021), Lucena et al. (2021), Novoa et al. (2021), Souza et al. (2022)

(Wei et al. 2021; Sun et al. 2022). This chapter explores this issue, discussing how natural and invaded ecosystem provide habitat and energy supply for the entire soil food web, how biological invasion changes habitat of the soil organisms, and two study cases considering invasive plant species (*Pinus* and *Acacia*) from tropical zones.

## 7.2   Soil Food Web in Moist and Dry Tropical Ecosystems

Tropical soils are ecologically rich, and these soils are considered a reservoir of much of the world's biodiversity that represent about 42% of the world's biomass carbon reserves (Jung et al. 2021; Paroshy et al. 2021). However, they are the most sensitive to human activities and ecological disturbances (Molotoks et al. 2018; Zhou et al. 2019; Sharma et al. 2020; Zhu et al. 2021a). In fact, human overpopulation and their activities have devastating effects on tropical soil overtime (Purswani et al. 2020; Makwinja et al. 2021). Some studies have reported that if the current rate of human activities continues (e.g., conventional agriculture, deforestation, fragmentation, forest fires, illegal hunting, infrastructure constructions, and wildlife trade), then many of the living organisms will vanish within 70 years (Scarano 2019; Scanes 2018; Edwards et al. 2019). To easier understand the tropical ecosystems and their effects on soil food web, we can split tropical soils in two main ecosystems:

1. Moist ecosystems: These regions are home to rain forests. They present high species diversity, high biomass production, and high litter input on soil surface (Giweta 2020; van Langenhove et al. 2020).
2. Dry ecosystems: These regions are home to dry forests and savannas. They support a fascinating diversity of living organisms' strategies to survive with fire, drought, and herbivory (Pennington et al. 2018; Amsten et al. 2021; Fill et al. 2021).

Moist tropical ecosystems are constantly changing due to land transformation, and the effects of abiotic entities on biodiversity and soil food web structure change more generally (Dietterich et al. 2021; Fibich et al. 2021). This kind of ecosystem exists on the extremes of temperature and precipitation and comprises a high biodiversity. Here, we can also find a complex soil food web. Moist ecosystems exist where high (sometimes moderate) rainfall combined with high temperatures is found. Compared to other ecosystems, moist ecosystem has high abundance of living organisms, which results in a robust soil food web. According to the studies done by Forstall-Sosa et al. (2020) and Melo et al. (2019), moist ecosystems present high abundance and high richness of soil organisms, and these variables are function of some abiotic and biotic properties overtime (Predictive models 1 and 2).

$$\text{Soil organisms' abundance} \sim \text{plant diversity} + \text{soil organic carbon} +$$
$$\text{root biomass} + \text{precipitation} + \text{time} \ (\text{Predictive model 1})$$

$$\text{Soil organisms' richness} \sim \text{litter deposition} + \text{litter quality} +$$
$$\text{shoot biomass} + \text{precipitation} + \text{time} \ (\text{Predictive model 2})$$

On the other hand, dry ecosystems contain highly fragmented forests (dry forests) and most of the arable land in the world (savannas) (Atsri et al. 2018; Janssen et al. 2018; Coelho et al. 2020). They receive less attention than moist ecosystem (rainforests), but they present moderate levels of biodiversity (Colli et al. 2020; Siyum 2020). In general, dry ecosystems biodiversity tends to be lower than in moist ecosystems (Raven et al. 2021). Dry ecosystems share a high proportion of endemism, but little is known about soil microbiota diversity (Álvarez-Lopeztello et al. 2019; Gnangui et al. 2021; Pompermaier et al. 2021). Considering the level of endemism, dry ecosystems contain proportionally more endemic species of plants and animals than moist ecosystems (Silva et al. 2019; Fernandes et al. 2020). This kind of ecosystem is constrained by low water availability, drought periods, and biological invasion. Abundance, richness, and diversity of soil organisms in such conditions are function of precipitation and soil organic carbon stocks as described by Souza and Santos (2018).

Both moist and dry ecosystems provide critical ecological services. For example, moist forests affect the entire global climate through primary producers' activity (photosynthesis) that influences carbon cycle and carbon sequestration (Djagbletey et al. 2018; Heinrich et al. 2021). Moist forests store up to 50% of all terrestrial carbon (Mitchard 2018; Schulte-Uebbing and de Vries 2018). On the other hand, dry forests comprise over 65% of the world's endemic species. For instance, habitat simplification, biological invasion, deforestation, mining, natural disturbances, and fire have already reduced soil biodiversity in such ecosystems (Pathak et al. 2019; Barfknecht et al. 2020; Verma et al. 2020). Although both moist and dry ecosystem could provide essential food resources by a wide range of soil organisms, they could lead to create different pathways to promote soil quality if we consider positive plant-soil feedback (Flores et al. 2020; Vasco-Palacios et al. 2020; D'Angioli et al. 2021).

At earlier stages, tropical ecosystems are characterized by seasonality (e.g., temperature and precipitation regimes) and their capacity to produce biomass and to stock carbon (Prach and Walker 2019; Bastida et al. 2021). Some studies have showed how the soil biota diversity changes as affected by seasonality in moist and dry ecosystems (Flores-Rentería et al. 2020; Heydari et al. 2020; Buscardo et al. 2021), and these findings could have been attributed to the following:

1. High litter inputs on soil surface and their residual effects as habitat and food resources to soil organisms are seasonal dependent.

2. Deposition, decomposition, and mineralization are the main soil organic matter traits that are modulated by thermal amplitude and rainfall in such ecosystems.
3. Habitat and energy provision for a wide range of soil organisms are dependent of plant community patterns.
4. Plant community composition that creates conditions for the soil food web establishment is soil and climate dependent.

## 7.3  Changing Habitat of Soil Organisms

Habitat can be defined as the natural environment (e.g., an array of resources, abiotic and biotic traits) of an organism that supports its survival and reproduction (Maclagan et al. 2018; Chytrý et al. 2019). Many young scientists in soil ecology occasionally try to define habitat as niche, but they have differences. Niche is the specific role (e.g., how does the organism contribute to the energy flow of the ecosystem?) that an organism plays in a natural or managed ecosystem, whereas habitat is the physical place where this organism lives and changes the environment as we can see in the eco-evo-devo field of science (Baedke and Mc Manus 2018; Gilbert 2021). In other words, habitat is the "address," and niche is the "function" (Halse 2018; Souza and Freitas 2017).

In natural ecosystems, we can find a very complex array of resources and abiotic and biotic traits that build specific community parameters such as richness, diversity, dominance, and abundance (Moreno et al. 2018; Staab and Schuldt 2020; Walters and Martiny 2020). Just seeing the diversity parameter helps us to understand community structure and its influence on the entire soil food web as described by the habitat provision hypothesis (Lennox et al. 2018; Davies et al. 2020). Some studies have reported in natural ecosystem the following phenomena:

1. Positive correlation with plant diversity and soil organism diversity (Bennett et al. 2020; Tresch et al. 2019; Marsden et al. 2020)
2. Positive correlation with plant diversity and abundance with litter quality and quantity, respectively (de Groote et al. 2018; Yang et al. 2019; Zhou et al. 2020)
3. Positive correlation with plant diversity with resources availability for soil organisms (Schmid et al. 2021; Zhang et al. 2022)
4. Positive correlation with plant diversity and abundance with soil biota activity (Heydari et al. 2020; da Silva et al. 2021)
5. Negative correlation with plant diversity with herbivory pressure (Weissflog et al. 2018; Muehleisen et al. 2020)
6. Negative correlation with plant diversity with allelopathic process (Gaggini et al. 2019; Kato-Noguchi and Kurniadie 2021; Wang et al. 2021b)

However, in managed ecosystems (e.g., monocropping farming systems, pasture, *Pinus* plantation, etc.), the human activity changes the habitat provision (Rodríguez-Romero et al. 2018; Powers and Jetz 2019). Considering that a habitat is a species-specific term and that every organism has particular habitat requirements, we must expect that habitat changes may directly affect soil organisms' community structure (Li et al. 2018; Wang et al. 2021b). When we change the plant community structure, for example, from a forest to a pasture, we can define it in soil ecology as habitat simplification (Hilpold et al. 2021; Paiva et al. 2020; Méndez-Rojas et al. 2021). It also may occur in monocropping systems that consider annual or perennial plant species (Zhao et al. 2018; Abán et al. 2021; Na et al. 2021). So, we can define habitat simplification as the process where there is a strong influence of the human activity that drastically reduces the biodiversity (in most of the cases, the first step is the plant diversity loss) in a specific area (Flores et al. 2018; Achury et al. 2020).

In the tropics, many studies have reported that the habitat simplification process has a negative influence on soil organisms' diversity and may create negative plant-soil feedback overtime (Pompermaier et al. 2020; Rai and Singh 2020). Both the soil quality and nutrient content hypothesis described by Melo et al. (2019) describe that plant species with high biomass production and the ability to establish symbiotic relationship with N-fixing bacteria can change soil organisms' community composition (e.g., richness, abundance, etc.). Some studies have reported that plant community composition can affect trophic structure into soil ecosystems as described by the enemies and resource concentration hypotheses (Gardarin et al. 2018; Staab and Schuldt 2020; Wan et al. 2020).

Considering the biological invasion promoted by invasive exotic plant species (IEPS), we must expect that changes in the habitat provision may have direct and indirect effects on soil organisms' community composition (Abgrall et al. 2018; Zhang et al. 2018; de Almeida et al. 2022). Overall, IEPS forms dense monodominance into the new ranges, thus drastically reducing the native plant species diversity and abundance (Adigbli et al. 2019; Spicer et al. 2022). We also must consider that the biological invasion can occur in several land uses (e.g., from a natural ecosystem to an unassisted forest restoration). The IEPS may affect the soil organisms' community composition in different ways, such as (1) favoring the process of functional redundancy (e.g., improving the spore's abundance from *Funneliformis mosseae* instead of the spore's abundance from *Claroideoglomus claroideum* in tropical acrisols); (2) changing the trophic structure of soil nematodes, thus affecting energy fluxes into the soil food web; (3) indirectly changing the soil organisms' richness, diversity, and dominance by reducing the richness and diversity of plant species; and (4) changing the soil traits by promoting its own seed bank (Fig. 7.1).

## 7.4   The Case of *Cryptostegia madagascariensis* and *Prosopis juliflora* in Northeastern Brazil

Invasive exotic plant species (IEPS) are considered the major threat for biodiversity around the world (Liu et al. 2019; Heringer et al. 2020). When introduced (accidently or not) outside their native range, IEPS are able to expand their range into natural areas, thus disrupting and outcompeting the native plant community

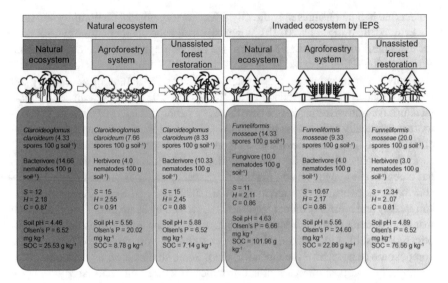

**Fig. 7.1** Influence of biological invasion by *Pinus* sp. on soil organisms' community structure. *S* species richness, *H* Shannon's diversity index, *C* Simpson's dominance index, *SOC* soil organic carbon. (Adapted from Laurindo et al. (2021), Souza et al. (2022))

(Hu et al. 2019; Fagúndez and Lema 2019; Weidlich et al. 2020). Usually, many studies have reported that an IEPS presents the following main characteristics:

1. They invest a lot of energy to produce viable seeds that are dispersed by wildlife and wind such as the cases of *Prosopis juliflora* and *Cryptostegia madagascariensis* in northeastern Brazil, respectively (Lucena et al. 2018; Nascimento et al. 2020).
2. There is lack of natural predators (no biological control of their population) that creates a monodominance of the IEPS as described by the natural enemy release hypothesis (Souza and Freitas 2017).
3. They present fast growth, and they can thrive on disturbances such as fire and drought (Seipel et al. 2018; Wood et al. 2019; Jambul et al. 2020).
4. They present longer growing seasons when compared with the native plant species (Hess et al. 2019; Staab et al. 2020).

The ecoregion located at northeastern Brazil (also called Caatinga ecoregion) has a rich and varied biodiversity (most of the soil organisms here are classified as endemic) and a long history of invasive plant species introductions which took place within governmental and environmental contexts, such as the cases of *P. juliflora* and *C. madagascariensis*, respectively (Pinto et al. 2020; Fulgêncio-Lima et al. 2021). The Caatinga ecoregion has a long history of biodiversity degradation (Tomassella et al. 2018; Vieira et al. 2020; de Oliveira et al. 2021). Just 1% of the total area from this ecoregion is placed in integral protection and conversation (e.g., National Parks from Chapada Diamantina, Serra da Capivara, and Serra das Confusões) (Antongiovanni et al. 2018; Dawson et al. 2021). The human activity has fragmented it, thus opening a window of opportunity to IEPS spread (Vardarman et al. 2018; Zhang et al. 2020).

*P. juliflora* is a highly IEPS in the Caatinga ecoregion (Mendonça et al. 2020; Dakhil et al. 2021). It covers nearly 37% of Caatinga territory (Souza et al. 2019; Pinto et al. 2020; Lucena et al. 2021). It is very tolerant to drought, fire disturbances, and herbivory pressure (Pandey et al. 2019; Arandhara et al. 2021). Some studies have reported that *P. juliflora* had been introduced in 1942 in the Brazilian territory. The Brazilian government aimed to introduce this IEPS as an economic alternative to regional agroforest system (da Silva and da Silva 2018; Carneiro-Junior et al. 2021). However, this IEPS started spreading its range in abandoned fields and degraded areas with several negative impacts on the entire ecosystem (Hussain et al. 2021; Choge et al. 2022). Increasing biological invasion by this exotic plant species is characterized by the invader removing competitively superior native plant species and increasing nutrient availability (Murugan et al. 2020; Elsheikh et al. 2021). Accordingly, to the study done by Souza et al. (2019), we must expect that (i) there is positive feedback between the symbiont communities and *P. juliflora*; (ii) *Prosopis juliflora* experienced positive feedback with specific symbionts, which indicates that this microbial community promotes the biological invasion processes in the Brazilian semi-arid region; (iii) native plant species does not perform well in a microbial community that has specific traits to colonize the IEPS roots; and (iv) the biological invasion promoted by *Prosopis juliflora* changed the outcome of symbiont interaction with native plant species, and the specific microbial community may promote the invasion success of the invader in field conditions.

On the other hand, the IEPS *C. madagascariensis* occupies dense areas in Ceara, Paraiba, and Rio Grande do Norte (Souza et al. 2018a, b; Pinto et al. 2020; Silveira et al. 2020). It occurs widespread and has destructive traits against the native vegetation, particularly *Mimosa tenuiflora*, *Copernicia prunifera*, and *Ziziphus joazeiro* because the exotic species can strangle and kill these native plant species by climbing over them and eliminating access to light (Alves and Fabricante 2019; Laurindo et al. 2021; Lucena et al. 2021), suggesting that these native plant species may be competitively affected by *C. madagascariensis* in the field conditions. *C. madagascariensis* is a woody perennial vine that is native to southwest Madagascar (Gracia et al. 2019; de Morais et al. 2021; Global Invasive Species Database 2020). It has been introduced in many Brazilian States in the 2010s by human activity because of its attractive flowers (da Silva et al. 2008). This IEPS forms dense aboveground stands, suggesting that it exerts a large impact on the entire ecosystem (Sousa et al. 2017; Vieira et al. 2019; Shiferaw et al. 2021).

Once introduced in the Caatinga ecoregion, both taxa have also dispersed within the new range through different pathways (Barros et al. 2021; Franco et al. 2021). They have influenced and impacted plant community composition, nutrient cycling, and symbiont interactions (Zhang et al. 2018; Souza et al. 2019; Chen et al. 2021). In the next subsection, we tried to show each impact promoted by *P. juliflora* and *C. madagascariensis* on each previously described abiotic and biotic trait.

### 7.4.1   Plant Community Composition

Overall, both *P. juliflora* and *C. madagascariensis* had showed negative impacts on plant community composition (e.g., richness, diversity, and abundance) (Fig. 7.2) in several ecosystems around northeastern Brazil (Nogueira Junior et al. 2019; Lucena et al. 2018; Fabricante et al. 2021). These IEPS have different strategies to spread themselves in the new range. The first one forms dense evergreen forest stands that present high growth rate, biomass production, and high N content in its litter (de Castro et al. 2020; Pik et al. 2020). It also has strong allelopathic effects that disrupt the soil seed bank and the natural regeneration (Abbas et al. 2019; Rai and Singh 2020; Zhu et al. 2021b), but there is evidence that shows positive correlation with *P. juliflora* (an IEPS), *Cereus jamacaru*, and *M. tenuiflora* (two native plant species from Brazilian northeast). On the other hand, the second one is a liana with low growth rate (Gracia et al. 2019; Lucena et al. 2019). It is very dependent to the native trees (e.g., it climbs up to the native tree canopy in search of sunlight) to its vertical support (Silveira et al. 2020; Souza et al. 2018a, b). According to Souza et al. (2016), there are two stages of biological invasion by *C. madagascariensis*. It is hard to detect negative impacts in the first stage that occurs in the first 3 years of its introduction on plant community composition. However, in the second stage, the negative impacts of *C. madagascariensis* are related to i) high biomass production, which reduces the access to sunlight for the native tree that the liana uses for vertical support, and ii) a hydrophobic litter, which creates a soil superficial sealing that decreases natural regeneration by killing the soil seed bank (Sousa et al. 2017; de Brito et al. 2021).

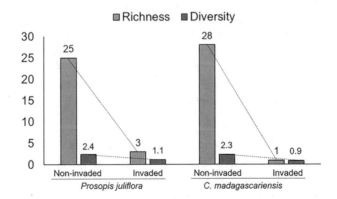

**Fig. 7.2** Plant community richness and diversity in invaded and non-invaded environments by *P. juliflora* and *C. madagascariensis*, respectively. (Adapted from Souza et al. (2019) and Lucena et al. (2021). Diversity is represented by Shannon's diversity index)

## 7.4.2   Nutrient Cycling

The IEPS can change litter decomposition ($k$ constant), litter deposition, soil organic matter, and litter P and N contents, as a strategy to their spread, growth, development, and reproduction (Braun et al. 2019; Zhang and Suseela 2021; Xu et al. 2022). The biological invasion process can disturb the continuous litter input into the invaded environment which promotes drastic changes in soil ecosystem (by creating new habitats and increasing energy supply to soil organisms). The invaders can easily increase the content of soil organic matter over time when compared to native plant species from semiarid and arid environments (Souza et al. 2016). Overall, the effects of biological invasion by *C. madagascariensis* and *P. juliflora* and their mechanisms underlying invasion success have been well-described around the world and illustrated by Lucena et al. (2018, 2021) and Souza et al. (2019). For example, *Prosopis juliflora* has shown near its rhizosphere and environment lower decay rate and soil organic matter content and higher litter deposition and litter P and N contents when compared to a rhizosphere from a native plant species. On the other hand, *C. madagascariensis* has shown higher decay rate, litter deposition, soil organic matter, and litter P and N contents near its rhizosphere when compared with the rhizosphere from a native plant species (Table 7.2).

It is expected that litter material with different composition may present dissimilar decomposition rate (e.g., high lignin content is related with low decomposition rate) (He et al. 2019; Qu et al. 2019). Some studies have reported faster litter decomposition in invaded environments when compared to non-invaded one (Yan et al. 2018; Jo et al. 2020; Kumar and Garkoti 2021). However, litter deposition, litter quality, and litter decomposition from invaded environments at tropical ecosystem are rare and poorly described. In the case of *P. juliflora* and *C. madagascariensis* invasion, the first one creates an island of fertility around itself, while the second one creates a physical barrier that negatively affects native soil seed bank (Salazar et al. 2019; Shiferaw et al. 2021; Lucena et al. 2021).

**Table 7.2**  Litter decay rate (k), litter deposition, soil organic matter, and nutrient contents in invaded and non-invaded environments by *P. juliflora* and *C. madagascariensis*, respectively

| IEPS | Environment | $k$ (years$^{-1}$) | Litter deposition (g m$^{-2}$) | Soil organic matter (g kg$^{-1}$) | Litter P content (g kg$^{-1}$) | Litter N content (g kg$^{-1}$) |
|---|---|---|---|---|---|---|
| *P. juliflora* | Invaded | 0.34 | 258.21 | 2.33 | 6.9 | 31.23 |
| | Non-invaded[a] | 0.94 | 98.32 | 4.56 | 2.3 | 18.98 |
| *C. madagascariensis* | Invaded | 2.01 | 537.23 | 8.45 | 2.9 | 20.05 |
| | Non-invaded | 0.94 | 103.28 | 4.78 | 2.3 | 19.75 |

[a]In the non-invaded environment, we used *Mimosa tenuiflora* as a model plant. (Adapted from Lucena et al. (2018, 2021) and Souza et al. (2019))

### 7.4.3　Symbionts: The Case of Arbuscular Mycorrhizal Fungi

Symbionts are some individuals from fungi and bacteria group that can establish a symbiotic relationship with plant species as we have previously described in this chapter. These soil organisms have an important role in increasing plant resistance against abiotic and biotic stresses (Guerrero-Galán et al. 2019; Diagne et al. 2020; Valliere et al. 2021; Souza et al. 2022). At tropical ecosystems, we can find arbuscular mycorrhizal fungi producing special structures (e.g., intraradical hyphae, vesicles, arbuscles, and auxiliary cells) inside roots. In invaded environments, symbionts improve the invader's growth, and there is evidence that shows a plant-host specificity between the IEPS and the arbuscular mycorrhizal fungi (AMF) (Awaydul et al. 2019; Zhang et al. 2019; Řezáčová et al. 2021; Sun et al. 2022). For example, in invaded environments by *P. juliflora* and *C. madagascariensis*, there is high dominance of *Funneliformis mosseae* and *Rhizophagus intraradices*, respectively (Souza et al. 2018a; Lucena et al. 2018; Akhtar et al. 2019). These AMF species have been reported in degraded environments. Also, we can find a drastic reduction on AMF richness and diversity (Fig. 7.3) in invaded environments by these two IEPS (Souza et al. 2016; Souza and Freitas 2017; Souza et al. 2019). Therefore, invaders seemed to be in advantage compared with natives with regard to emergence rate, survival rate, plant growth, or nutrient uptake by profiting from beneficial AMF species (e.g., *Funneliformis mosseae*, *Rhizophagus intraradices*, and *Claroideoglomus claroideum*) (Zhang et al. 2018; Souza et al. 2018b; Chen et al. 2021). The differences in AMF community structure between natives and invasive plant species were revealed by the decreased AMF species richness in all invasive plant species root zone.

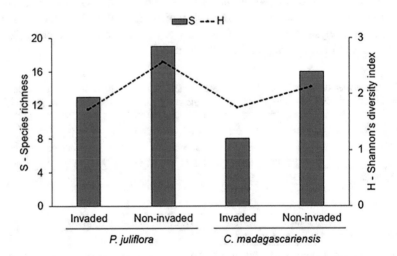

**Fig. 7.3** Arbuscular mycorrhizal fungal richness (S) and diversity (H) in invaded and non-invaded environments by *P. juliflora* and *C. madagascariensis*, respectively. (Adapted from Souza et al. (2016), Souza and Freitas (2017) and Souza et al. (2019))

We must hypothesize that three different mechanisms may be involved in the AMF species richness and diversity reduction. First, IEPS form large monospecific stands that reduce the diversity of native plant species that might be available as host plants to the AMF community (see the host specificity hypothesis proposed by Souza et al. (2019)). Consequently, this changes soil organic carbon inputs (Mbaabu et al. 2021; McLeod et al. 2021), and monospecific stands of IEPS result in a reduction of mycorrhizal plant communities that negatively impact the AMF growth and their asymbiotic phase (Caravaca et al. 2020; Laurindo et al. 2021; Rai 2021). Second, metabolites (e.g., allelopathic compounds) produced by invaders negatively affect native plant growth by disrupting their mutualistic associations with the native AMF community (Qin and Yu 2019; Simberloff et al. 2021). Overall, IEPS can produce metabolites that are novel for the native AMF community in their introduced areas, and these secondary compounds directly limit AMF growth, spore germination, and root colonization (e.g., both asymbiotic and symbiotic phases) (Pinzone et al. 2018; Zubek et al. 2022). Consequently, the most beneficial AMF from the native AMF community is favored, while the growth of the less favorable ones is inhibited. Finally, the introduction of IEPS might cause changes in soil chemical properties that may indirectly affect AMF community composition and thus contribute to a successful establishment and spread of the invader (Liu et al. 2021; Stefanowicz et al. 2021).

## 7.5 Conclusion

We may summarize the following aspects: (1) natural environments provide habitat and energy supply for a wide range of soil organisms and (2) invasive plant species alter plant community composition, soil chemical properties, and AMF community composition. Scientific evidence suggests that the invaded environment can facilitate the establishment and subsequent spread of invasive in the tropics.

## References

Abán CL, Brandan CP, Verdenelli R, Huidobro J, Meriles JM, Gil SV (2021) Changes in microbial and physicochemical properties under cover crop inclusion in a degraded common bean monoculture system. Eur J Soil Biol 107:103365. https://doi.org/10.1016/j.ejsobi.2021.103365

Abbas AM, Figueroa ME, Castillo JM (2019) Burial effects on seed germination and seedling establishment of *Prosopis juliflora* (SW.) DC. Arid Land Res Manag 33:55–69. https://doi.org/10.1080/15324982.2018.1457103

Abgrall C, Forey E, Mignot L, Chauvat M (2018) Invasion by Fallopia japonica alters soil food webs through secondary metabolites. Soil Biol Biochem 127:100–109. https://doi.org/10.1016/j.soilbio.2018.09.016

Abgrall C, Forey E, Chauvat M (2019) Soil fauna responses to invasive alien plants are determined by trophic groups and habitat structure: a global meta-analysis. Oikos 128(10):1390–1401. https://doi.org/10.1111/oik.06493

Achury R, de Ulloa PC, Arcila Á, Suarez AV (2020) Habitat disturbance modifies dominance, coexistence, and competitive interactions in tropical ant communities. Ecol Entomol 45(6):1247–1262. https://doi.org/10.1111/een.12908

Adigbli DM, Anning AK, Adomako JK, Fosu-Mensah (2019) Effects of *Broussonetia papyrifera* invasion and land use on vegetation characteristics in a tropical forest of Ghana. J For Res 30:1363–1373. https://doi.org/10.1007/s11676-018-0691-9

Akhtar O, Mishra R, Kehri HK (2019) Arbuscular mycorrhizal association contributes to Cr accumulation and tolerance in plants growing on Cr contaminated soils. Proc Natl Acad Sci India Sect B Biol Sci 89:63–70. https://doi.org/10.1007/s40011-017-0914-4

Álvarez-Lopeztello J, del Castillo RF, Robles C, Hernández-Cuevas LV (2019) Spore diversity of arbuscular mycorrhizal fungi in human-modified neotropical ecosystems. Ecol Res 34(3):394–405. https://doi.org/10.1111/1440-1703.12004

Alves JS, Fabricante JR (2019) Exotic invasive flora evaluation on different environments and preservation conditions from a caatinga area, Petrolina, PE. Gaia Scientia 13(1). https://doi.org/10.22478/ufpb.1981-1268.2019v13n1.39803

Amsten K, Cromsigt JP, Kuijper DP, Loberg JM, Churski M, Niklasson M (2021) Fire-and herbivory-driven consumer control in a savanna-like temperate wood-pasture: an experimental approach. J Ecol 109(12):4103–4114. https://doi.org/10.1111/1365-2745.13783

Antongiovanni M, Venticinque EM, Fonseca CR (2018) Fragmentation patterns of the Caatinga drylands. Landsc Ecol 33:1353–1367. https://doi.org/10.1007/s10980-018-0672-6

Arandhara S, Sathishkumar S, Gupta S, Baskaran N (2021) Influence of invasive *Prosopis juliflora* on the distribution and ecology of native blackbuck in protected areas of Tamil Nadu. India Eur J Wildl Res 6:48. https://doi.org/10.1007/s10344-021-01485-3

Atsri HK, Konko Y, Cuni-Sanchez A, Abotsi KE, Kokou K (2018) Changes in the West African forest-savanna mosaic, insights from central Togo. PLoS One 13(10):e0203999. https://doi.org/10.1371/journal.pone.0203999

Awaydul A, Zhu W, Yuan Y, Xiao J, Hu H, Chen X, Koide RT, Cheng L (2019) Common mycorrhizal networks influence the distribution of mineral nutrients between an invasive plant, *Solidago canadensis*, and a native plant, *Kummerowa striata*. Mycorrhiza 29:29–38. https://doi.org/10.1007/s00572-018-0873-5

Baedke J, Mc Manus SF (2018) From seconds to eons: time scales, hierarchies, and processes in evo-devo. Stud Hist Phil Scie Part C: Stud Hist Phil Biol Biomed Sci 72:38–48. https://doi.org/10.1016/j.shpsc.2018.10.006

Barfknecht DF, Li G, Martinez KA, Gibson DJ (2020) Interactive disturbances drive community composition, heterogeneity, and the niches of invasive exotic plant species during secondary succession. Plant Ecol Divers 13(5–6):363–375. https://doi.org/10.1080/17550874.2020.1841313

Barros V, Oliveira MT, Santos MG (2021) Low foliar construction cost and strong investment in root biomass in *Calotropis procera*, an invasive species under drought and recovery. Flora 280:151848. https://doi.org/10.1016/j.flora.2021.151848

Baruch EM, Bateman HL, Lytle DA, Merritt DM, Sabo JL (2021) Integrated ecosystems: linking food webs through reciprocal resource reliance. Ecol Soc Am 102(9):e03450. https://doi.org/10.1002/ecy.3450

Bastida F, Eldridge DJ, García C, Png GK, Bardgett RD, Delgado-Baquerizo M (2021) Soil microbial diversity–biomass relationships are driven by soil carbon content across global biomes. ISME J 15:2081–2091. https://doi.org/10.1038/s41396-021-00906-0

Battini N, Giachetti CB, Castro KL, Bortolus A, Schwindt E (2021) Predator–prey interactions as key drivers for the invasion success of a potentially neurotoxic sea slug. Biol Invasions 23:1207–1229. https://doi.org/10.1007/s10530-020-02431-1

Bempah AN, Kyereh B, Ansong M, Asante W (2021) The impacts of invasive trees on the structure and composition of tropical forests show some consistent patterns but many are context dependent. Biol Invasions 23:1307–1319. https://doi.org/10.1007/s10530-020-02442-y

Bennett JA, Koch AM, Forsythe J, Johnson NC, Tilman D, Klironomos J (2020) Resistance of soil biota and plant growth to disturbance increases with plant diversity. Ecol Lett 23(1):119–128. https://doi.org/10.1111/ele.13408

Bertelsmeier C (2021) Globalization and the anthropogenic spread of invasive social insects. Curr Opin Insect Sci 46:16–23. https://doi.org/10.1016/j.cois.2021.01.006

Braun K, Collantes MB, Yahdjian L, Escartin C, Anchorena JA (2019) Increased litter decomposition rates of exotic invasive species *Hieracium pilosella* (Asteraceae) in southern Patagonia, Argentina. Plant Ecol 220:393–403. https://doi.org/10.1007/s11258-019-00922-3

Brito SF, Pinheiro CL, Matos DMS, Medeiros Filho S (2021) Establishment of *Cryptostegia madagascariensis* in the semiarid: what is the role of abiotic factors in germination and initial growth? Scientia Plena 17(5):052401–052401. https://doi.org/10.14808/sci.plena.2021.052401

Buscardo E, Souza RC, Meir P, Geml J, Schmidt SK, da Costa ACL, Nagy L (2021) Effects of natural and experimental drought on soil fungi and biogeochemistry in an Amazon rain forest. Commun Earth Environ 2:55. https://doi.org/10.1038/s43247-021-00124-8

Caravaca F, Rodríguez-Caballero G, Campoy M, Sanleandro PM, Roldán A (2020) The invasion of semiarid Mediterranean sites by *Nicotiana glauca* mediates temporary changes in mycorrhizal associations and a permanent decrease in rhizosphere activity. Plant Soil 450:217–229. https://doi.org/10.1007/s11104-020-04497-1

Carneiro-Junior JAM, de Oliveira GF, Alves CT, Andrade HMC, Vieira B, de Melo SA, Torres EA (2021) Valorization of *Prosopis juliflora* woody biomass in northeast Brazilian through dry torrefaction. Energies 14(12):3465. https://doi.org/10.3390/en14123465

Cazetta AL, Zenni RD (2020) Pine invasion decreases density and changes native tree communities in woodland Cerrado. Plant Ecol Diver 13:85–91. https://doi.org/10.1080/17550874.2019.1675097

Chen Q, Wu WW, Qi SS, Cheng H, Li Q, Ran Q, Dai SC, Du DL, Egan S, Thomas T (2021) Arbuscular mycorrhizal fungi improve the growth and disease resistance of the invasive plant *Wedelia trilobata*. J Appl Microbiol 130(2):582–591. https://doi.org/10.1111/jam.14415

Choge S, Mbaabu PR, Muturi GM (2022) Management and control of the invasive *Prosopis juliflora* tree species in Africa with a focus on Kenya. In: Puppo MC, Fleker P (eds) *Prosopis* as a heat tolerant nitrogen fixing desert food legume. Elsevier, Academic Press. https://doi.org/10.1016/C2020-0-00062-9

Chytrý M, Hájek M, Kočí M, Pešout P, Roleček J, Sádlo J, Šumberová K, Sychra J, Boublík K, Douda J, Grulich V, Händrij H, Hédl R, Lustyk P, Navrátilová J, Novák P, Peterka T, Vydrová A, Chobot K (2019) Red list of habitats of the Czech Republic. Ecol Indic 106:105446. https://doi.org/10.1016/j.ecolind.2019.105446

Coelho AJP, Magnago LFS, Matos FAR, Mota NM, Diniz ES, Meira-Neto JAA (2020) Effects of anthropogenic disturbances on biodiversity and biomass stock of Cerrado, the Brazilian savanna. Biodivers Conserv 29:3151–3168. https://doi.org/10.1007/s10531-020-02013-6

Colli GR, Vieira CR, Dianese JC (2020) Biodiversity and conservation of the Cerrado: recent advances and old challenges. Biodivers Conserv 29:1465–1475. https://doi.org/10.1007/s10531-020-01967-x

D'Angioli AM, Dantas VL, Lambais M, Meir P, Oliveira RS (2021) No evidence of positive feedback between litter deposition and seedling growth rate in Neotropical savannas. Plant Soil 469:305–320. https://doi.org/10.1007/s11104-021-05163-w

da Silva VDA, da Silva AMM, e Silva JHC, Cotas SL (2018) Neurotoxicity of *Prosopis juliflora*: from natural poisoning to mechanism of action of its piperidine alkaloids. Neurotox Res 34:878–888. https://doi.org/10.1007/s12640-017-9862-2

Da Silva JL, Barreto RW, Pereira OL (2008) Pseudocercospora *cryptostegiae-madagascariensis* sp. nov. on *Cryptostegia madagascariensis,* an exotic vine involved in major biological invasions in Northeast Brazil. Mycopathologia 166(2):87–91. https://doi.org/10.1007/s11046-008-9120-5

Da Silva JVCL, Ferris H, Cares JE, Esteves AM (2021) Effect of land use and seasonality on nematode faunal structure and ecosystem functions in the Caatinga dry forest. Eur J Soil Biol 103:103296. https://doi.org/10.1016/j.ejsobi.2021.103296

Dairel M, Fidelis A (2020) The presence of invasive grasses affects the soil seed bank composition and dynamics of both invaded and non-invaded areas of open savannas. J Environ Manag 276:111291. https://doi.org/10.1016/j.jenvman.2020.111291

Dakhil MA, El-Keblawy A, El-Sheikh MA, Halmy MWA, Ksiksi T, Hassan WA (2021) Global invasion risk assessment of *Prosopis juliflora* at biome level: does soil matter? Biology 10(3):203. https://doi.org/10.3390/biology10030203

Davies RW, Edwards DP, Edwards FA (2020) Secondary tropical forests recover dung beetle functional diversity and trait composition. Anim Conserv 23(5):617–627. https://doi.org/10.1111/acv.12584

Dawson N, Carvalho WD, Bezerra JS, Todeschini F, Tabarelli M, Mustin K (2021) Protected areas and the neglected contribution of indigenous peoples and local communities: struggles for environmental justice in the Caatinga dry forest. People Nat. https://doi.org/10.1002/pan3.10288

de Almeida T, Forey E, Chauvat M (2022) Alien invasive plant effect on soil fauna is habitat dependent. Diversity 14(2):61. https://doi.org/10.3390/d14020061

de Castro WAC, Almeida RV, Xavier RO, Bianchini I, Moya H, Matos DMS (2020) Litter accumulation and biomass dynamics in riparian zones in tropical South America of the Asian invasive plant *Hedychium coronarium* J. König (Zingiberaceae). Plant Ecol Divers 13:47–59. https://doi.org/10.1080/17550874.2019.1673496

de Groote SRE, Vanhellemont M, Baeten L, de Schrijver A, Bonte D, Lens L, Verheyen K (2018) Tree species diversity indirectly affects nutrient cycling through the shrub layer and its high-quality litter. Plant Soil 427:335–350. https://doi.org/10.1007/s11104-018-3654-1

De Long JR, Fry EL, Veen GF, Kardol P (2019) Why are plant–soil feedbacks so unpredictable, and what to do about it? Funct Ecol 33(1):118–128. https://doi.org/10.1111/1365-2435.13232

de Morais SMD, Pinheiro HB, Rebouças Filho JV, Cavalcante GS, Bonilla OH (2021) *Cryptostegia* genus: phytochemistry, biologial activities and industrial applications. Química Nova 44:709–716. https://doi.org/10.21577/0100-4042.20170716

de Oliveira ML, dos Santos CA, de Oliveira G, Perez-Marin AM, Santos CA (2021) Effects of human-induced land degradation on water and carbon fluxes in two different Brazilian dryland soil covers. Sci Total Environ 792:148458. https://doi.org/10.1016/j.scitotenv.2021.148458

Delon N, Purves D (2018) Wild animal suffering is intractable. J Agric Environ Ethics 31:239–260. https://doi.org/10.1007/s10806-018-9722-y

Diagne N, Ngom M, Djighaly PI, Fall D, Hocher V, Svistoonoff S (2020) Roles of Arbuscular Mycorrhizal fungi on plant growth and performance: importance in biotic and abiotic stressed regulation. Diversity 12(10):370. https://doi.org/10.3390/d12100370

Dietterich LH, Karpman J, Neupane A, Chiochina M, Cusack DF (2021) Carbon content of soil fractions varies with season, rainfall, and soil fertility across a lowland tropical moist forest gradient. Biogeochemistry 155:431–452. https://doi.org/10.1007/s10533-021-00836-1

Djagbletey ED, Logah V, Ewusi-Mensah N, Tuffour HO (2018) Carbon stocks in the Guinea savanna of Ghana: estimates from three protected areas. Biotropica 50(2):225–233. https://doi.org/10.1111/btp.12529

Edwards DP, Socolar JB, Mills SC, Burivalova Z, Koh LP, Wilcove DS (2019) Conservation of tropical forests in the anthropocene. Curr Biol 29(19):R1008–R1020. https://doi.org/10.1016/j.cub.2019.08.026

Elsheikh EAE, El-Keblawy A, Mosa KA, Okoh AI, Saadoun I (2021) Role of endophytes and rhizosphere microbes in promoting the invasion of exotic plants in arid and semi-arid areas: a review. Sustainability 13(23):13081. https://doi.org/10.3390/su132313081

Eshete A, Treydte AC, Hailemariam M, Solomon N, Dejene T, Yilma Z, Birhane E (2020) Variations in soil properties and native woody plant species abundance under *Prosopis juliflora* invasion in Afar grazing lands. Ethiopia Ecol Proc 9:36. https://doi.org/10.1186/s13717-020-00240-x

Fabricante JR, de Araújo KCT, Almeida TS, Santos JPB, Reis DO (2021) Invasive alien plants in Sergipe, northeastern Brazil. Neotrop Biol Conserv 16(1):89–104. https://doi.org/10.3897/neotropical.16.e56427

Fagúndez J, Lema MA (2019) A competition experiment of an invasive alien grass and two native species: are functionally similar species better competitors? Biol Invasions 21:3619–3631. https://doi.org/10.1007/s10530-019-02073-y

Ferlian O, Eisenhauer N, Aguirrebengoa M, Camara M, Ramirez-Rojas I, Santos F, Tanalgo K, Thakur MP (2018) Invasive earthworms erode soil biodiversity: a meta-analysis. J Anim Ecol 87(1):162–172. https://doi.org/10.1111/1365-2656.12746

Fernandes MF, Cardoso D, de Queiroz LP (2020) An updated plant checklist of the Brazilian Caatinga seasonally dry forests and woodlands reveals high species richness and endemism. J Arid Environ 174:104079. https://doi.org/10.1016/j.jaridenv.2019.104079

Fibich P, Novotný V, Ediriweera S, Gunatilleke S, Gunatilleke N, Molem K, Weiblen GD, Lepš J (2021) Common spatial patterns of trees in various tropical forests: small trees are associated with increased diversity at small spatial scales. Ecol Evol 11(12):8085–8095. https://doi.org/10.1002/ece3.7640

Fill JM, Zamora C, Baruzzi C, Salazar-Castro J, Crandall RM (2021) Wiregrass (Aristida beyrichiana) survival and reproduction after fire in a long-unburned pine savanna. PLoS One 16(2):e0247159. https://doi.org/10.1371/journal.pone.0247159

Flores LMA, Zanette LRS, Araujo FS (2018) Effects of habitat simplification on assemblages of cavity nesting bees and wasps in a semiarid neotropical conservation area. Biodivers Conserv 27:311–328. https://doi.org/10.1007/s10531-017-1436-3

Flores BM, Oliveira RS, Rowland L, Quesada CA, Lambers H (2020) Editorial special issue: plant-soil interactions in the Amazon rainforest. Plant Soil 450:1–9. https://doi.org/10.1007/s11104-020-04544-x

Flores-Rentería D, Sánchez-Gallén I, Morales-Rojas D, Larsen J, Álvarez-Sánchez J (2020) Changes in the abundance and composition of a microbial community associated with land use change in a Mexican tropical rain Forest. J Soil Sci Plant Nutr 20:1144–1155. https://doi.org/10.1007/s42729-020-00200-6

Forstall-Sosa KS, Souza TAF, Lucena EO, Silva SIA, Ferreira JTA, Silva TN, Santos D, Niemeyer JC (2020) Soil macroarthropod community and soil biological quality index in a green manure farming system of the Brazilian semi-arid. Biologia 76:907–917. https://doi.org/10.2478/s11756-020-00602-y

Franco JR, Paterno GB, Ganade G (2021) The influence of herbaceous vegetation on the colonization of native and invasive trees: consequences for semiarid forest restoration, Restorat Ecol:e13595. https://doi.org/10.1111/rec.13595

Frelich LE, Blossey B, Cameron EK, Dávalos A, Eisenhauer N, Fahey T, Ferlian O, Groffman PM, Larson E, Loss SR, Maerz JC, Nuzzo V, Yoo K, Reich PB (2019) Side-swiped: ecological cascades emanating from earthworm invasions. Front Ecol Environ 17(9):502–510. https://doi.org/10.1002/fee.2099

Fulgêncio-Lima LG, Andrade AFA, Vilela B, Lima-Júnior DP, de Souza RA, Sgarbi LF, Simião-Ferreira J, de Marco JP, Silva DP (2021) Invasive plants in Brazil: climate change effects and detection of suitable areas within conservation units. Biol Invasions 23(5):1577–1594. https://doi.org/10.1007/s10530-021-02460-4

Gaggini L, Rusterholz HP, Baur B (2019) The invasion of an annual exotic plant species affects the above-and belowground plant diversity in deciduous forests to a different extent. Perspect Plant Ecol Evol Syst 38:74–83. https://doi.org/10.1016/j.ppees.2019.04.004

Gao L, Hou B, Cai ML, Zhai JJ, Li WH, Peng CL (2018) General laws of biological invasion based on the sampling of invasive plants in China and the United States. Glob Ecol Conserv 16:e00448. https://doi.org/10.1016/j.gecco.2018.e00448

Gardarin A, Plantegenest M, Bischoff A, Valatin-Morison M (2018) Understanding plant–arthropod interactions in multitrophic communities to improve conservation biological control: useful traits and metrics. J Pest Sci 91:943–955. https://doi.org/10.1007/s10340-018-0958-0

Gilbert SF (2021) Evolutionary developmental biology and sustainability: a biology of resilience. Evol Dev 23(4):273–291. https://doi.org/10.1111/ede.12366

Giweta M (2020) Role of litter production and its decomposition, and factors affecting the processes in a tropical forest ecosystem: a review. J Ecol Environ 44:11. https://doi.org/10.1186/s41610-020-0151-2

Global Invasive Species Database (GISD) (2020) Species profile *Cryptostegia madagascariensis*. http://www.iucngisd.org/gisd/species.php?sc=1628. Accessed 12 Feb 2022

Gnangui SLE, Fossou RK, Ebou A, Amon CER, Koua DK, Kouadjo CGZ, Cowan DA, Zézé A (2021) The rhizobial microbiome from the tropical savannah zones in northern Côte d'Ivoire. Microorganisms 9(9):1842. https://doi.org/10.3390/microorganisms9091842

Gracia A, Rangel-Buitrago N, Castro-Barros JD (2019) Non-native plant species in the Atlantico department coastal dune systems, Caribbean of Colombia: a new management challenge. Mar Pollut Bull 141:603–610. https://doi.org/10.1016/j.marpolbul.2019.03.009

Guerrero-Galán C, Calvo-Polanco M, Zimmermann SD (2019) Ectomycorrhizal symbiosis helps plants to challenge salt stress conditions. Mycorrhiza 29:291–301. https://doi.org/10.1007/s00572-019-00894-2

Halse SA (2018) Subterranean fauna of the arid zone. In: Lambers H (ed) On the ecology of Australia's arid zone. Springer, Cham. https://doi.org/10.1007/978-3-319-93943-8_9

He M, Zhao R, Tian Q, Huang L, Wang X, Liu F (2019) Predominant effects of litter chemistry on lignin degradation in the early stage of leaf litter decomposition. Plant Soil 442:453–469. https://doi.org/10.1007/s11104-019-04207-6

Heinrich VH, Dalagnol R, Cassol HL, Rosan TM, de Almeida CT, Silva Junior CH, Campanharo WA, House JI, Sitch S, Jales TC, Adami M, Anderson LO, Aragão LE (2021) Large carbon sink potential of secondary forests in the Brazilian Amazon to mitigate climate change. Nat Commun 12(1):1–11. https://doi.org/10.1038/s41467-021-22050-1

Heringer G, Thiele J, Meira-Neto JAA, Neri AV (2019) Biological invasion threatens the sandy-savanna *Mussununga* ecosystem in the Brazilian Atlantic Forest. Biol Invasions 21:2045–2057. https://doi.org/10.1007/s10530-019-01955-5

Heringer G, Thiele J, do Amaral CH, JAA M-N, FAR M, JRK L, Buttschardt TK, Neri AV (2020) Acacia invasion is facilitated by landscape permeability: the role of habitat degradation and road networks. Appl Veg Sci 23(4):598–609. https://doi.org/10.1111/avsc.12520

Hess MCM, Mesléard F, Buisson E (2019) Priority effects: emerging principles for invasive plant species management. Ecol Eng 127:48–57. https://doi.org/10.1016/j.ecoleng.2018.11.011

Heydari M, Eslaminejad P, Kakhki FV, Mirab-Balou M, Omidipour R, Prévosto B, Koock Y, Lucas-Borja ME (2020) Soil quality and mesofauna diversity relationship are modulated by woody species and seasonality in semiarid oak forest. For Ecol Manag 473:118332. https://doi.org/10.1016/j.foreco.2020.118332

Hilpold A, Seeber J, Fontana V, Niedrist G, Rief A, Steinwandter M, Tasser E, Tappeiner U (2021) Decline of rare and specialist species across multiple taxonomic groups after grassland intensification and abandonment. Biodivers Conserv 27:3729–3744. https://doi.org/10.1007/s10531-018-1623-x

Hu CC, Lei YB, Tan YH, Sun XC, Xu H, Liu CQ, Liu XY (2019) Plant nitrogen and phosphorus utilization under invasive pressure in a montane ecosystem of tropical China. J Ecol 107(1):372–386. https://doi.org/10.1111/1365-2745.13008

Hulme PE (2020) Plant invasions in New Zealand: global lessons in prevention, eradication and control. Biol Invasions 22:1539–1562. https://doi.org/10.1007/s10530-020-02224-6

Hussain MI, Shackleton R, El-Keblawy A, González L, Trigo MM (2021) Impact of the invasive *Prosopis juliflora* on terrestrial ecosystems. In: Lichtfouse E (ed) Sustainable agriculture reviews 52. Sustainable agriculture reviews, vol 52. Springer, Cham. https://doi.org/10.1007/978-3-030-73245-5_7

Jambul R, Limin A, Ali AN, Slik F (2020) Invasive *Acacia mangium* dominance as an indicator for heath forest disturbance. Environ Sustainabil Indicat 8:100059. https://doi.org/10.1016/j.indic.2020.100059

Janssen TA, Ametsitsi GK, Collins M, Adu-Bredu S, Oliveras I, Mitchard ET, Veenendaal EM (2018) Extending the baseline of tropical dry forest loss in Ghana (1984–2015) reveals drivers of major deforestation inside a protected area. Biol Conserv 218:163–172. https://doi.org/10.1016/j.biocon.2017.12.004

Jo I, Fridley JD, Frank DA (2020) Rapid leaf litter decomposition of deciduous understory shrubs and lianas mediated by mesofauna. Plant Ecol 221:63–68. https://doi.org/10.1007/s11258-019-00992-3

Jung M, Arnell A, de Lamo X, García-Rangel S, Lewis M, Mark J, Merow C, Miles L, Ondo I, Pironon S, Ravilious C, Rivers M, Schepashenko D, Tallowin O, van Soesbergen A, Govaersts BBL, Enquist BJ, Feng X, Gallagher R, Maitner B, Meiri S, Mullingan M, Ofer G, Roll U, Hanson JO, Jetz W, di Marco M, McGowan J, Rinnan DS, Sachs JD, Lesiv M, Adams VM, Andrew SC, Burger JR, Hannah L, Marquet PA, McCarthy JK, Morueta-Holme N, Newman EA, Park DS, Roehrdanz PR, Svenning J, Violle C, Wieringa JJ, Wynne G, Fritz S, Strassburg BBN, Obersteiner M, Kapos V, Nurgess N, Schmidt-Traub G, Pironon S (2021) Areas of global importance for conserving terrestrial biodiversity, carbon and water. Nat EcolEvolution 5:1499–1509. https://doi.org/10.1038/s41559-021-01528-7

Kareiva P, Carranza V (2018) Existential risk due to ecosystem collapse: nature strikes back. Futures 102:39–50. https://doi.org/10.1016/j.futures.2018.01.001

Kato-Noguchi H, Kurniadie D (2021) Allelopathy of *Lantana camara* as an invasive plant. Plan Theory 10(5):1028. https://doi.org/10.3390/plants10051028

Kumar M, Garkoti SC (2021) Functional traits, growth patterns, and litter dynamics of invasive alien and co-occurring native shrub species of chir pine forest in the central Himalaya, India. Plant Ecol 222:723–735. https://doi.org/10.1007/s11258-021-01140-6

Laurindo LK, Souza TAF, da Silva LJR, Casal TB, Pires KJC, Kormann S, Schmitt DE, Siminski A (2021) Arbuscular mycorrhizal fungal community assembly in agroforestry systems from the Southern Brazil. Biologia 76:1099–1107. https://doi.org/10.1007/s11756-021-00700-5

Lennox GD, Gardner TA, Thomson JR, Ferreira J, Berenguer E, Lees AC, Nally RM, Aragão LEOC, Ferraz SFB, Louzada J, Moura NG, Oliveira VHF, Pardini R, Solar RRC, Vaz-de-Mello FZ, Vieira ICG, Barlow J (2018) Second rate or a second chance? Assessing biomass and biodiversity recovery in regenerating Amazonian forests. Glob Chang Biol 24(12):5680–5694. https://doi.org/10.1111/gcb.14443

Li FR, Liu JL, Ren W, Liu LL (2018) Land-use change alters patterns of soil biodiversity in arid lands of northwestern China. Plant Soil 428:371–388. https://doi.org/10.1007/s11104-018-3673-y

Linders TEW, Schaffner U, Eschen R, Abebe A, Choge SK, Nigatu L, Mbaabu PR, Shiferaw H, Allan E (2019) Direct and indirect effects of invasive species: biodiversity loss is a major mechanism by which an invasive tree affects ecosystem functioning. J Ecol 107(6):2660–2672. https://doi.org/10.1111/1365-2745.13268

Liu X, Blackburn TM, Song T, Li X, Huang C, Li Y (2019) Risks of biological invasion on the belt and road. Curr Biol 29(3):499–505. https://doi.org/10.1016/j.cub.2018.12.036

Liu G, Liu RL, Zhang WG, Yang Y, Bi Z, Li M, Chen X, Nie H, Zhu Z (2021) Arbuscular mycorrhizal colonization rate of an exotic plant, *Galinsoga quadriradiata*, in mountain ranges changes with altitude. Mycorrhiza 31:161–171. https://doi.org/10.1007/s00572-020-01009-y

Lucena EO, Souza TAF, Araújo JS, Aandrade LA, Santos D, Podestá GS (2018) Occurrence and distribution of *Gigaspora* under *Cryptostegia madagascariensis* Bojer Ex Decne in Brazilian tropical seasonal dry forest. Rev Agropec Tec 39:221–227. https://doi.org/10.25066/agrotec.v39i3.40055

Lucena EO, Souza T, da Silva SIA, Kormann S, da Silva LJR, Laurindo LK, Forstall-Sossa KS, de Andrade LA (2021) Soil biota community composition as affected by *Cryptostegia madagascariensis* invasion in a tropical Cambisol from North-Eastern Brazil. Trop Ecol 62(4):662–669. https://doi.org/10.1007/s42965-021-00177-y

Maclagan SJ, Coates T, Ritchie EG (2018) Don't judge habitat on its novelty: assessing the value of novel habitats for an endangered mammal in a peri-urban landscape. Biol Conserv 223:11–18. https://doi.org/10.1016/j.biocon.2018.04.022

Makwinja R, Kaunda E, Mengistou S, Alamirew T (2021) Impact of land use/land cover dynamics on ecosystem service value—a case from Lake Malombe. Southern Malawi Environ Monit Assess 193(8):1–23. https://doi.org/10.1080/27658511.2021.1969139

Marsden C, Martin-Chave A, Cortet J, Hedde M, Capowiez Y (2020) How agroforestry systems influence soil fauna and their functions – a review. Plant Soil 453:29–44. https://doi.org/10.1007/s11104-019-04322-4

Maxwell SL, Cazalis V, Dudley N, Hoffmann M, Rodrigues AS, Stolton S, Visconti P, Woodley S, Kingston N, Luís E, Maron M, Strassburg BBN, Wnger A, Jonas HD, Venter O, Watson JE (2020) Area-based conservation in the twenty-first century. Nature 586(7828):217–227. https://doi.org/10.1038/s41586-020-2773-z

Mbaabu PR, Olago D, Gichaba M, Eckert S, Eschen R, Oriaso S, Coge SK, Linders TEW, Schaffner U (2021) Restoration of degraded grasslands, but not invasion by *Prosopis juliflora,* avoids trade-offs between climate change mitigation and other ecosystem services. Sci Rep 10:20391. https://doi.org/10.1038/s41598-020-77126-7

McLeod ML, Bullington L, Cleveland CC, Rousk J, Lekberg Y (2021) Invasive plant-derived dissolved organic matter alters microbial communities and carbon cycling in soils. Soil Biol Biochem 156:108191. https://doi.org/10.1016/j.soilbio.2021.108191

Melo LN, Souza TAF, Santos D (2019) Cover crop farming system affects macroarthropods community diversity in Regosol of Caatinga, Brazil. Biologia 74:1653–1660. https://doi.org/10.2478/s11756-019-00272-5

Méndez-Rojas DM, Cultid-Medina C, Escobar F (2021) Influence of land use change on rove beetle diversity: a systematic review and global meta-analysis of a mega-diverse insect group. Ecol Indic 122:107239. https://doi.org/10.1016/j.ecolind.2020.107239

Mendonça MF, Pedroso PM, Pimentel LA, Madureira KM, Macêdo J, D'Soares CS, Silva AWO, Peixoto TC (2020) Epidemiological aspects of natural poisoning by Prosopis juliflora in ruminants in semiarid areas of the state of Bahia, Brazil, invaded by the plant. Pesquisa Veterinária Brasileira 40:501–513. https://doi.org/10.1590/1678-5150-PVB-6664

Milanović M, Knapp S, Pyšek P, Kühn I (2020) Linking traits of invasive plants with ecosystem services and disservices. Ecosyst Serv 42:101072. https://doi.org/10.1016/j.ecoser.2020.101072

Mitchard ETA (2018) The tropical forest carbon cycle and climate change. Nature 559:527–534. https://doi.org/10.1038/s41586-018-0300-2

Molotoks A, Stehfest E, Doelman J, Albanito F, Fitton N, Dawson TP, Smith P (2018) Global projections of future cropland expansion to 2050 and direct impacts on biodiversity and carbon storage. Glob Chang Biol 24(12):5895–5908. https://doi.org/10.1111/gcb.14459

Mora FA (2018) Spatial framework for detecting anthropogenic impacts on predator-prey interactions that sustain ecological integrity in Mexico. Ecol Process 7:35. https://doi.org/10.1186/s13717-018-0146-4

Moreno CE, Calderón-Patrón JM, Martín-Regalado N, Martínez-Falcón AP, Ortega-Martínez IJ, Rios-Díaz CL, Rosas F (2018) Measuring species diversity in the tropics: a review of methodological approaches and framework for future studies. Biotropica 50(6):929–941. https://doi.org/10.1111/btp.12607

Muehleisen AJ, Engelbrecht BM, Jones FA, Manzané-Pinzón E, Comita LS (2020) Local adaptation to herbivory within tropical tree species along a rainfall gradient. Ecology 101(11):e03151. https://doi.org/10.1002/ecy.3151

Murugan R, Beggi F, Prabakaran N, Maqsood S, Joergensen RG (2020) Changes in plant community and soil ecological indicators in response to *Prosopis juliflora* and *Acacia mearnsii* invasion and removal in two biodiversity hotspots in Southern India. Soil Ecol Lett 2:61–72. https://doi.org/10.1007/s42832-019-0020-z

Na X, Ma S, Ma C, Liu Z, Xu P, Zhu H, Liang W, Kardol P (2021) *Lycium barbarum* L. (goji berry) monocropping causes microbial diversity loss and induces *Fusarium* spp. enrichment at distinct soil layers. Appl Soil Ecol 168:104107. https://doi.org/10.1016/j.apsoil.2021.104107

Nascimento CES, da Silva CAD, Leal IR, de Souza WT, Serrão JE, Zanuncio JC, Tabarelli M (2020) Seed germination and early seedling survival of the invasive species *Prosopis juliflora* (Fabaceae) depend on habitat and seed dispersal mode in the Caatinga dry forest. PeerJ 8:e9607. https://doi.org/10.7717/peerj.9607

Nogueira Junior FC, Pagotto MA, Aragão JRV, Roig FA, Ribeiro AS, Lisi CS (2019) The hydrological performance of *Prosopis juliflora* (Sw.) growth as an invasive alien tree species in the semiarid tropics of northeastern Brazil. Biol Invasions 21:2561–2575. https://doi.org/10.1007/s10530-019-01994-y

Novoa A, Foxcroft LC, Keet JH, Pyšek P, Le Roux JJ (2021) The invasive cactus *Opuntia stricta* creates fertility islands in African savannas and benefits from those created by native trees. Sci Rep 11:20748. https://doi.org/10.1038/s41598-021-99857-x

Paiva IG, Auad AM, Veríssimo BA, Silveira LCP (2020) Differences in the insect fauna associated to a monocultural pasture and a silvopasture in Southeastern Brazil. Sci Rep 10:12112. https://doi.org/10.1038/s41598-020-68973-5

Pandey CB, Singh AK, Saha D, Mathur BK, Tewari JC, Kumar M, Goyal RK, Mathur M, Gaur MK (2019) *Prosopis juliflora* (Swartz) DC.: an invasive alien in community grazing lands and its control through utilization in the Indian Thar Desert. Arid Land Res Manag 33(4):427–448. https://doi.org/10.1080/15324982.2018.1564402

Paroshy NJ, Doraisami M, Kish R, Martin AR (2021) Carbon concentration in the world's trees across climatic gradients. New Phytol 232(1):123–133. https://doi.org/10.1111/nph.17587

Pathak R, Negi VS, Rawal RS, Bhatt ID (2019) Alien plant invasion in the Indian Himalayan region: state of knowledge and research priorities. Biodivers Conserv 28:3073–3102. https://doi.org/10.1007/s10531-019-01829-1

Pennington RT, Lehmann CE, Rowland LM (2018) Tropical savannas and dry forests. Curr Biol 28(9):R541–R545. https://doi.org/10.1016/j.cub.2018.03.014

Pik D, Lucero JE, Lortie CJ, Braun J (2020) Light intensity and seed density differentially affect the establishment, survival, and biomass of an exotic invader and three species of native competitors. Community Ecol 21:259–272. https://doi.org/10.1007/s42974-020-00027-2

Pinto AS, Monteiro FKS, Ramos MB, Araújo RCC, Lopes SF (2020) Invasive plants in the Brazilian Caatinga: a scientometric analysis with prospects for conservation. Neotrop Biol Conserv 15(4):503–520. https://doi.org/10.3897/neotropical.15.e57403

Pinzone P, Potts D, Pettibone G, Warren R II (2018) Do novel weapons that degrade mycorrhizal mutualisms promote species invasion? Plant Ecol 219:539–548. https://doi.org/10.1007/s11258-018-0816-4

Pompermaier VT, Kisaka TB, Ribeiro JF, Nardoto GB (2020) Impact of exotic pastures on epigeic arthropod diversity and contribution of native and exotic plant sources to their diet in the central Brazilian Savanna. Pedobiologia 78:150607. https://doi.org/10.1016/j.pedobi.2019.150607

Pompermaier VT, Campani AR, Dourado E, Coletta LD, Bustamante MMDC, Nardoto GB (2021) Soil mesofauna drives litter decomposition under combined nitrogen and phosphorus additions in a Brazilian woodland savanna. Austral Ecol. https://doi.org/10.1111/aec.13082

Powers RP, Jetz W (2019) Global habitat loss and extinction risk of terrestrial vertebrates under future land-use-change scenarios. Nat Clim Chang 9:323–329. https://doi.org/10.1038/s41558-019-0406-z

Prach K, Walker LR (2019) Differences between primary and secondary plant succession among biomes of the world. J Ecol 107(2):510–516. https://doi.org/10.1111/1365-2745.13078

Purswani E, Pathak B, Kumar D, Verma S (2020) Land-use change as a disturbance regime. In: Shukla V, Kumar N (eds) Environmental concerns and sustainable development. Springer, Singapore. https://doi.org/10.1007/978-981-13-6358-0_6

Pyšek P, Hulme PE, Simberloff D, Bacher S, Blackburn TM, Carlton JT, Dawson W, Essl F, Foxcroft LC, Genovesi P, Jeschke JM, Kühn I, Liebhold AM, Mandrak NE, Meyerson LA, Pauchard A, Pergl J, Roy HE, Seebens H, van Kleunen M, Vilà M, Wingfield MJ, Richardson DM (2020) Scientists' warning on invasive alien species. Biol Rev, 95: 1511-1534. https://doi.org/10.1111/brv.12627

Qin F, Yu S (2019) Arbuscular mycorrhizal fungi protect native woody species from novel weapons. Plant Soil 440:39–52. https://doi.org/10.1007/s11104-019-04063-4

Qu H, Pan C, Zhao X, Lian J, Wang S, Wang X, Ma X, iu L (2019) Initial lignin content as an indicator for predicting leaf litter decomposition and the mixed effects of two perennial gramineous plants in a desert steppe: a 5-year long-term study. Land Degrad Dev 30(14): 1645–1654. https://doi.org/10.1002/ldr.3343

Rai PK (2021) Environmental degradation by invasive alien plants in the anthropocene: challenges and prospects for sustainable restoration. Sci, Anthr. https://doi.org/10.1007/s44177-021-00004-y

Rai PK, Singh JS (2020) Invasive alien plant species: their impact on environment, ecosystem services and human health. Ecol Indic 111:106020. https://doi.org/10.1016/j.ecolind.2019.106020

Raven PH, Gereau RE, Phillipson PB, Chatelain C, Jenkins CN, Ulloa CU (2021) The distribution of biodiversity richness in the tropics. Sci Adv 6:37. https://doi.org/10.1126/sciadv.abc6228

Renčo M, Jurová J, Gömöryová E, Čerevková A (2021) Long-term Giant hogweed invasion contributes to the structural changes of soil nematofauna. Plan Theory 10(10):2103. https://doi.org/10.1016/j.gecco.2021.e01470

Řezáčová V, Řezáč M, Gryndler M, Hršelová H, Gryndlerova H, Michalova T (2021) Plant invasion alters community structure and decreases diversity of arbuscular mycorrhizal fungal communities. Appl Soil Ecol 167:104039. https://doi.org/10.1016/j.apsoil.2021.104039

Rodríguez-Caballero G, Caravaca F, Díaz G, Torres P, Roldán A (2020) The invader *Carpobrotus edulis* promotes a specific rhizosphere microbiome across globally distributed coastal ecosystems. Sci Total Environ 719:137347. https://doi.org/10.1016/j.scitotenv.2020.137347

Rodríguez-Romero AJ, Rico-Sánchez AE, Mendoza-Martínez E, Gómez-Ruiz A, Sedeño-Díaz JE, López-López E (2018) Impact of changes of land use on water quality, from tropical Forest to anthropogenic occupation: a multivariate approach. Water 10(11):1518. https://doi.org/10.3390/w10111518

Salazar PC, Navarro-Cerrillo RM, Grados N, Cruz G, Barrón V, Villar R (2019) Tree size and leaf traits determine the fertility island effect in *Prosopis pallida* dryland forest in Northern Peru. Plant Soil 437:117–135. https://doi.org/10.1007/s11104-019-03965-7

Scanes CG (2018) Human activity and habitat loss: destruction, fragmentation, and degradation. In: Scanes CG, Toukhsati SR (eds). Animals and Human Society. https://doi.org/10.1016/C2014-0-03860-9

Scarano FR (2019) Biodiversity sector: risks of temperature increase to biodiversity and ecosystems. In: Nobre C, Marengo J, Soares W (eds) Climate change risks in Brazil. Springer, Cham. https://doi.org/10.1007/978-3-319-92881-4_5

Schmid MW, van Moorsel SJ, Hahl T, De Luca E, De Deyn GB, Wagg C, Niklaus PA, Schimid B (2021) Effects of plant community history, soil legacy and plant diversity on soil microbial communities. J Ecol 109(8):3007–3023. https://doi.org/10.1111/1365-2745.13714

Schulte-Uebbing L, de Vries W (2018) Global-scale impacts of nitrogen deposition on tree carbon sequestration in tropical, temperate, and boreal forests: a meta-analysis. Glob Chang Biol 24(2):e416–e431. https://doi.org/10.1111/gcb.13862

Seeney A, Eastwood S, Pattison Z, Willby NJ, Bull CD (2020) All change at the water's edge: invasion by non-native riparian plants negatively impacts terrestrial invertebrates. Biol Invasions 21:1933–1946. https://doi.org/10.1007/s10530-019-01947-5

Seipel T, Rew LJ, Taylor KT, Maxwell BD, Lehnhoff EA (2018) Disturbance type influences plant community resilience and resistance to *Bromus tectorum* invasion in the sagebrush steppe. Appl Veg Sci 21(3):385–394. https://doi.org/10.1111/avsc.12370

Shackleton RT, Larson BM, Novoa A, Richardson DM, Kull CA (2019) The human and social dimensions of invasion science and management. J Environ Manag 229:1–9. https://doi.org/10.1016/j.jenvman.2018.08.041

Sharma S, MacKenzie RA, Tieng T, Soben K, Tulyasuwan N, Resanond A, Blate G, Creighton M, Litton CM (2020) The impacts of degradation, deforestation and restoration on mangrove ecosystem carbon stocks across Cambodia. *Sci Total Environ* 706:135416. https://doi.org/10.1016/j.scitotenv.2019.135416

Shiferaw W, Demissew S, Bekele T, Aynekulu E, Pitroff W (2021) Invasion of *Prosopis juliflora* and its effects on soil physicochemical properties in Afar region, Northeast Ethiopia. Int Soil Water Conserv Res 9(4):631–638. https://doi.org/10.1016/j.iswcr.2021.04.003

Siddiqui A, Bamisile BS, Khan MM, Islam W, Hafeez M, Bodlah I, Xu Y (2021) Impact of invasive ant species on native fauna across similar habitats under global environmental changes. Environ Sci Pollut Res 28:54362–54382. https://doi.org/10.1007/s11356-021-15961-5

Silva JPM, da Silva MLM, da Silva EF, da Silva GF, de Mendonca AR, Cabacinha CD, Araújo EF, Santos JS, Vieira GC, de Almeida MNF, Fernandes MRM (2019) Computational techniques applied to volume and biomass estimation of trees in Brazilian savanna. J Environ Manag 249:109368. https://doi.org/10.1016/j.jenvman.2019.109368

Silveira AP, Menezes BSD, Loiola MIB, Lima-Verde LW, Zanina DN, Carvalho ECDD, Souza BC, da Costa RC, Mantovani W, de Menezes MOT, Flores LMA, FCB N, Matias LQ, Barbosa LS, Gomes FM, Cordeiro LS, Sampaio VS, MEP B, Soares Neto LC, da Silva MAP, Campos NB, de Oliveira AA, FSD A (2020) Flora and annual distribution of flowers and fruits in the Ubajara National Park, Ceará. Brazil Floresta e Ambiente 27. https://doi.org/10.1590/2179-8087.005819

Simberloff D, Kaur H, Kalisz S, Bezemer TM (2021) Novel chemicals engender myriad invasion mechanisms. New Phytol 232(3):1184–1200. https://doi.org/10.1111/nph.17685

Siyum ZG (2020) Tropical dry forest dynamics in the context of climate change: syntheses of drivers, gaps, and management perspectives. Ecol Process 9:25. https://doi.org/10.1186/s13717-020-00229-6

Sousa FQ, Andrade LA, Silva PCC, Souza BCQ, Xavier KRF (2017) Banco de sementes do solo de caatinga invadida por *Cryptostegia madagascariensis* Bojer ex Decne. Revista Brasileira de Ciências Agrárias 12:220–226. https://doi.org/10.5039/agraria.v12i2a5440

Souza TAF, Freitas H (2017) Arbuscular mycorrhizal fungal community assembly in the Brazilian tropical seasonal dry forest. Ecol Proc. https://doi.org/10.1186/s13717-017-0072-x

Souza TAF, Santos D (2018) Effects of using different host plants and long-term fertilization systems on population sizes of infective arbuscular mycorrhizal fungi. Symbiosis 76:139–149. https://doi.org/10.1007/s13199-018-0546-3

Souza TAF, Rodriguez-Echeverria S, Andrade LA, Freitas H (2016) Could biological invasion by *Cryptostegia madagascariensis* alter the composition of the arbuscular mycorrhizal fungal community in semi-arid Brazil? Acta Bot Bras. https://doi.org/10.1590/0102-33062015abb0190

Souza TAF, de Andrade LA, Freitas H, Sandim AS (2018a) Biological invasion influences the outcome of plant-soil feedback in the invasive plant species from the Brazilian semi-arid. Microbioal Ecology 76:102–112. https://doi.org/10.1007/s00248-017-0999-6

Souza TAF, Rodriguez-Echeverria S, Freitas H, de Andrade LA (2018b) *Funneliformis mosseae* and invasion by exotic legumes in a Brazilian tropical seasonal dry Forest. Russ J Ecol 49:500–506. https://doi.org/10.1134/S1067413618060127

Souza TAF, Santos D, Andrade LA, Freitas H (2019) Plant-soil feedback of two legumes species in semi-arid Brazil. Braz J Microbiol 50:1011–1020. https://doi.org/10.1007/s42770-019-00125-y

Souza TAF, Barros IC, da Silva LJR, Laurindo LK, Nascimento GD, Lucena EO, Martins M, dos Santos VB (2022) Comunidade da microbiota do solo reunida em espécies vegetais nativas da Amazônia legal brasileira. Simbiose. https://doi.org/10.1007/s13199-021-00828-7

Spicer ME, Radhamoni HVN, Duguid MC, Queenborough SA, Comita LS (2022) Herbaceous plant diversity in forest ecosystems: patterns, mechanisms, and threats. Plant Ecol 223:117–129. https://doi.org/10.1007/s11258-021-01202-9

Staab M, Schuldt A (2020) The influence of tree diversity on natural enemies—a review of the "enemies" hypothesis in forests. Curr Forestry Rep 6:243–259. https://doi.org/10.1007/s40725-020-00123-6

Staab M, Pereira-Peixoto MH, Klein AM (2020) Exotic garden plants partly substitute for native plants as resources for pollinators when native plants become seasonally scarce. Oecologia 194:465–480. https://doi.org/10.1007/s00442-020-04785-8

Stefanowicz AM, Majewska ML, Stanek M, Nobis M, Zubek S (2018) Differential influence of four invasive plant species on soil physicochemical properties in a pot experiment. J Soils Sediments 18:1409–1423. https://doi.org/10.1007/s11368-017-1873-3

Stefanowicz AM, Stanek M, Majewska ML, Nobis M, Zubek S (2019) Invasive plant species identity affects soil microbial communities in a mesocosm experiment. Appl Soil Ecol 136:168–177. https://doi.org/10.1016/j.apsoil.2019.01.004

Stefanowicz AM, Kapusta P, Stanek M, Frąc M, Oszust K, Woch MW, Zubek S (2021) Invasive plant *Reynoutria japonica* produces large amounts of phenolic compounds and reduces the biomass but not activity of soil microbial communities. Sci Total Environ 767:145439

Sun F, Ou Q, Yu H, Li N, Peng C (2019) The invasive plant *Mikania micrantha* affects the soil foodweb and plant-soil nutrient contents in orchards. Soil Biol Biochem 139:107630. https://doi.org/10.1016/j.soilbio.2019.107630

Sun F, Zeng L, Cai M, Chauvat M, Forey E, Tariq A, Graciano C, Zhang Z, Gu Y, Zeng F, Gong Y, Wang F, Wang M (2022) An invasive and native plant differ in their effects on the soil food-web and plant-soil phosphorus cycle. Geoderma 410:115672. https://doi.org/10.1016/j.geoderma.2021.115672

Thapa S, Chitale V, Rijal SJ, Bisht N, Shrestha BB (2018) Understanding the dynamics in distribution of invasive alien plant species under predicted climate change in Western Himalaya. PLoS One 13(4):e0195752. https://doi.org/10.1371/journal.pone.0195752

Tomassella J, Vieira RMSP, Barbosa AA, Rodriguez DA, Santana MO, Sestini MF (2018) Desertification trends in the ortheast of Brazil over the period 2000–2016. Int J Appl Earth Obs Geoinf 73:197–206. https://doi.org/10.1016/j.jag.2018.06.012

Torres N, Herrera I, Fajardo L, Bustamante RO (2021) Meta-análise do impacto de invasões de plantas nas comunidades microbianas do solo. BMC Eco Evo 21:172. https://doi.org/10.1186/s12862-021-01899-2

Tresch S, Frey D, Le Bayon RC, Zanetta A, Rasche F, Fliessbach A, Moretti M (2019) Litter decomposition driven by soil fauna, plant diversity and soil management in urban gardens. Sci Total Environ 658:1614–1629. https://doi.org/10.1016/j.scitotenv.2018.12.235

Tsujimoto M, Kajikawa Y, Tomita J, Matsumoto Y (2018) A review of the ecosystem concept—towards coherent ecosystem design. Technol Forecast Soc Chang 136:49–58. https://doi.org/10.1016/j.techfore.2017.06.032

Uboni C, Tordoni E, Brandmayr P, Battistella S, Bragato G, Castello M, Colombetta G, Poldini L, Bacaro G (2019) Exploring cross-taxon congruence between carabid beetles (Coleoptera: Carabidae) and vascular plants in sites invaded by *Ailanthus altissima* versus non-invaded sites: the explicative power of biotic and abiotic factors. Ecol Indic 103:145–155. https://doi.org/10.1016/j.ecolind.2019.03.052

Valliere JM, D'Agui HM, Dixon KW, Nevill PG, Wong WS, Zhong H, Veneklaas EJ (2021) Stockpiling disrupts the biological integrity of topsoil for ecological restoration. Plant Soil 1(2):e12027. https://doi.org/10.1002/2688-8319.12027

Van Langenhove L, Verryckt LT, Bréchet L, Courtois EA, Stahl C, Hofhansl F, Bauters M, Sardans J, Boeckx P, Fransen E, Peñuelas J, Janssens IA (2020) Atmospheric deposition of elements and its relevance for nutrient budgets of tropical forests. Biogeochemistry 149:175–193. https://doi.org/10.1007/s10533-020-00673-8

Vardarman J, Berchová-Bímová K, Pěknicová J (2018) The role of protected area zoning in invasive plant management. Biodivers Conserv 27:1811–1829. https://doi.org/10.1007/s10531-018-1508-z

Vasco-Palacios AM, Bahram M, Boekhout T, Tedersoo L (2020) Carbon content and pH as important drivers of fungal community structure in three Amazon forests. Plant Soil. https://doi.org/10.1007/s11104-019-04218-3

Verma AK, Rout PR, Lee E, Bhunia P, Bae J, Surampalli RY, Zhang TC, Tyagu RD, Lin P, Chen Y (2020) Biodiversity and sustainability. In: Surampolli R, Zhang T, Goyal MK, Brar S, Tyagi R (eds) Sustainability: fundamentals and applications. Wiley, New Jersey. https://doi.org/10.1002/9781119434016.ch12

Vieira EA, Galvão FCA, Barros AL (2019) Influence of water limitation on the competitive interaction between two Cerrado species and the invasive grass *Brachiaria brizantha* cv. Piatã Plant Physiol Biochem 135:206–214. https://doi.org/10.1016/j.plaphy.2018.12.002

Vieira RMDSP, Sestini MF, Tomasella J, Marchezini V, Pereira GR, Barbosa AA, Santos FC, Rodriguez DA, do Nascimento FR, Santana MO, FCB C, JOHB O (2020) Characterizing spatio-temporal patterns of social vulnerability to droughts, degradation and desertification in the Brazilian northeast. Environ Sustainabil Indicat 5:100016. https://doi.org/10.1016/j.indic.2019.100016

Walters KE, Martiny JBH (2020) Alpha-, beta-, and gamma-diversity of bacteria varies across habitats. PLoS One 15(9):e0233872. https://doi.org/10.1371/journal.pone.0233872

Wan NF, Zheng XR, Fu LW, Kiær LP, Zhang Z, Chaplin-Kramer R, Dainese M, Tan J, Qiu S, Hu Y, Tian W, Nie W, Ju R, Deng J, Jiang J, Cai Y Li B (2020) Global synthesis of effects of plant species diversity on trophic groups and interactions. Nat Plants 6: 503–510. https://doi.org/10.1038/s41477-020-0654-y

Wang W, Feng C, Liu F, Li J (2020) Biodiversity conservation in China: a review of recent studies and practices. Environ Sci Ecotechnol 2:100025. https://doi.org/10.1016/j.ese.2020.100025

Wang C, Masoudi A, Wang M, Yang J, Yu Z, Liu J (2021a) Land-use types shape soil microbial compositions under rapid urbanization in the Xiong'an new area. China Sci Total Environ 777:145976. https://doi.org/10.1016/j.scitotenv.2021.145976

Wang A, Melton AE, Soltis DE, Soltis PS (2021b) Potential distributional shifts in North America of allelopathic invasive plant species under climate change models. Plant Divers. https://doi.org/10.1016/j.pld.2021.06.010

Wei Q, Yin R, Huang J, Vogler AP, Li Y, Miao X, Kardol P (2021) The diversity of soil mesofauna declines after bamboo invasion in subtropical China. Sci Total Environ 789:147982. https://doi.org/10.1016/j.scitotenv.2021.147982

Weidlich EW, Flórido FG, Sorrini TB, Brancalion PH (2020) Controlling invasive plant species in ecological restoration: a global review. J Appl Ecol 57(9):1806–1817. https://doi.org/10.1111/1365-2664.13656

Weissflog A, Markesteijn L, Lewis OT, Comita LS, Engelbrecht BM (2018) Contrasting patterns of insect herbivory and predation pressure across a tropical rainfall gradient. Biotropica 50(2):302–311. https://doi.org/10.1111/btp.12513

Wood DJA, Seipel T, Irvine KM, Rew LJ, Stoy PC (2019) Fire and development influences on sagebrush community plant groups across a climate gradient in northern Nevada. Ecosphere 20(12):e02990. https://doi.org/10.1002/ecs2.2990

Xu H, Liu Q, Wang S, Yang G, Xue S (2022) A global meta-analysis of the impacts of exotic plant species invasion on plant diversity and soil properties. Sci Total Environ 810:152286. https://doi.org/10.1016/j.scitotenv.2021.152286

Yan J, Wang L, Hu Y, Tsang YF, Zhang Y, Wu J, Fu X, Sun Y (2018) Plant litter composition selects different soil microbial structures and in turn drives different litter decomposition pattern and soil carbon sequestration capability. Geoderma 319:194–203. https://doi.org/10.1016/j.geoderma.2018.01.009

Yang X, Qu YB, Yang N, Zhao H, Wang JL, Zhao NX, Gao YB (2019) Litter species diversity is more important than genotypic diversity of dominant grass species *Stipa grandis* in influencing litter decomposition in a bare field. Sci Total Environ 666:490–498. https://doi.org/10.1016/j.scitotenv.2019.02.247

Yletyinen J, Perry GLW, Burge OR, Mason NWH, Stahlmann-Brown P (2021) Invasion landscapes as social-ecological systems: role of social factors in invasive plant species control. People Nat 3(4):795–810. https://doi.org/10.1002/pan3.10217

Zhang Z, Suseela V (2021) Nitrogen availability modulates the impacts of plant invasion on the chemical composition of soil organic matter. Soil Biol Biochem 156:108195. https://doi.org/10.1016/j.soilbio.2021.108195

Zhang F, Li Q, Yerger EH, Chen X, Shi Q, Wan F (2018) AM fungi facilitate the competitive growth of two invasive plant species, *Ambrosia artemisiifolia* and *Bidens pilosa*. Mycorrhiza 28:703–715. https://doi.org/10.1007/s00572-018-0866-4

Zhang P, Li B, Wu J, Hu S (2019) Invasive plants differentially affect soil biota through litter and rhizosphere pathways: a meta-analysis. Ecol Lett 22(1):200–210. https://doi.org/10.1111/ele.13181

Zhang X, Wei H, Zhao Z, Liu J, Zhang Q, Zhang X, Gu W (2020) The global potential distribution of invasive plants: *Anredera cordifolia* under climate change and human activity based on random forest models. Sustainability 12(4):1491. https://doi.org/10.3390/su12041491

Zhang Y, Peng S, Chen X, Chen HY (2022) Plant diversity increases the abundance and diversity of soil fauna: a meta-analysis. Geoderma 411:115694. https://doi.org/10.1016/j.geoderma.2022.115694

Zhao Q, Xiong W, Xing Y, Sun Y, Lin X, Dong Y (2018) Long-term coffee monoculture alters soil chemical properties and microbial communities. Sci Rep 8:6116. https://doi.org/10.1038/s41598-018-24537-2

Zhou Y, Hartemink AE, Shi Z, Liang Z, Lu Y (2019) Land use and climate change effects on soil organic carbon in North and Northeast China. Sci Total Environ 647:1230–1238. https://doi.org/10.1016/j.scitotenv.2018.08.016

Zhou S, Butenschoen O, Barantal S, Handa IT, Makkonen M, Vos V, Aerts R, Berg MP, McKie PB, Van Ruijven J, Hättensnchwiler S, Scheu S (2020) Decomposition of leaf litter mixtures across biomes: the role of litter identity, diversity and soil fauna. J Ecol 108(6):2283–2297. https://doi.org/10.1111/1365-2745.13452

Zhu G, Qiu D, Zhang Z, Sang L, Liu Y, Wang L, Zhao K, Ma H, Xu Y, Wan Q (2021a) Land-use changes lead to a decrease in carbon storage in arid region. China Ecol Indicat 127:107770. https://doi.org/10.1016/j.ecolind.2021.107770

Zhu X, Yi Y, Huang L, Zhang C, Shao H (2021b) Metabolomics reveals the allelopathic potential of the invasive plant *Eupatorium adenophorum*. Plan Theory 10(7):1473. https://doi.org/10.3390/plants10071473

Zubek S, Kapusta P, Stanek M, Woch MW, Błaszkowski J, Stefanowicz AM (2022) *Reynoutria japonica* invasion negatively affects arbuscular mycorrhizal fungi communities regardless of the season and soil conditions. Appl Soil Ecol 169:104152. https://doi.org/10.1016/j.apsoil.2021.104152

# Chapter 8
# Natural Disasters

**Abstract** In this chapter, the natural disaster concept is introduced as the main influencer of soil ecosystem structure and functioning. At the tropics, there are several natural disaster types, but in this chapter our aim is to present some related aspects of fire events, landslides, and hurricanes and their influence on soil abiotic and biotic traits. These natural disasters are formed through the human activities, urban growth, nature of soils (geology, morphology, and topography), and climate conditions. In this chapter, we focused only on describing the most important and significant effects of natural disasters on soil ecosystem, such as plant community structure, soil organic matter, and N:P stoichiometry. In view of this, it is important that soil ecologist must consider the characterization of both soil abiotic and biotic traits.

**Keywords** Fire · Hurricanes · Landslides · Soil ecosystem · Soil food web

**Questions Covered in the Chapter**

1. How can a natural disaster alter the soil ecosystem?
2. How frequent are natural disasters at the tropics?
3. What is the importance to consider the fire effect on the soil organisms' community composition and soil nutrient loss?
4. What are the main impacts of landslides on soil food web?
5. How do the hurricanes alter soil food web?

## 8.1 Introduction

Soil ecosystem is a multiphase system that covers the Earth's surface, and it presents specific abiotic and biotic traits at the tropics (Slimane and El-hafid 2021). Soil geology, morphology, and topography must be considered to predict predisposed areas for natural disasters (Jacobs et al. 2017; Morgado et al. 2018). Fire events, landslides, and hurricanes have specific ranges for their occurrence, and the human activity must act as a catalyzer for their occurrence at both temporal and spatial scale (Keefer 1994; Ponge 2015). Natural disasters mainly influence the soil ecosystem by disrupting and disturbing several soil functions (Lindenmayer et al. 2017). They also may permanently affect soil cover and vegetation restoration (Coyle et al. 2017). As a result, the soil ecosystem at the tropics is strongly influenced by these natural events that are becoming more frequent because of the climate change (Nunes et al. 2021).

Natural disasters can change both habitat and energy provision, and in turn, they create the main negative plant-soil feedback by affecting soil physical, chemical, and biological properties (De Deyn and Kooistra 2021). Indeed, the natural disaster is considered the main threat from the tropics (Bieng et al. 2021), and the soil ecologist needs to understand their influence on the soil matrix (Vogel et al. 2021). Some studies have pointed the effect of natural disasters on soil abiotic traits, but studies considering their influence on soil biotic traits are rare and scares (Yunus et al. 2021). Here, the soil ecologist will find a background on the main effects of fire events, landslides, and hurricanes on plant community composition, soil organic matter, and N:P stoichiometry.

## 8.2   Burning the Living Soil: The Case of Fire in West-Central Brazil

Fire events in west-central Brazil (where the Pantanal took place – the world's largest wetland) have been reported every mid-July since the 1960s, but these last 10 years, they become more massive because of climate change (Kogan 2019). The massive fire affects the entire ecosystem, burning trees, injuring, and killing animals and destroying the soil ecosystem (Chandra and Bhardwaj 2015). The Pantanal is a magnet for organisms' diversity because its characteristics (abundant water) and location (it is located between the Amazon rainforest and the Paraguay's dry forest) and the fire events are threatening this Brazilian ecoregion (Junk et al. 2006a). A biodiversity loss near to 35% of invertebrates (most of them endemic) is expected every year after fire events (Junk et al. 2006b; Barlow et al. 2018).

Overall, the fire events occur on deforested lands (Cochrane 2003). It emphasizes the importance to avoid the deforestation at the tropics (Austin et al. 2019). We can split fire events based on nature and impact range in three categories:

1. Natural fire events: It is related with the drought periods when fire burns the desiccated vegetation (Fig. 8.1). Some plant species need to breakup their seeds dormancy and to improve natural regeneration. This event has local impact and, in some biotic traits, has a neutral impact on the soil ecosystem (Haberle et al. 2001).

**Fig. 8.1** Natural fire events during dry periods. The presence of grasses and desiccated vegetation facilitates the fire intensity and its spread

2. Agricultural purposes: It is related with the use of flames by farmers to cheaply return nutrients to the soil and renew their crop systems. It takes place in extensive areas and increases soil erosion, nutrient leaching, and soil biodiversity loss overtime. It has a negative regional impact. Unfortunately, this kind of fire event has a historic force across the Pantanal (Berlinck et al. 2022).
3. Random fire events: It is related with areas surrounding cities, roads, and waste deposits. It has a negative local impact by destroying urban structure, blocking roads, and producing toxic smoke that can affect human and animals' health (Manrique-Pineda et al. 2021).

In the field of soil ecology, we must consider that fire events have direct impact on net primary production, atmospheric chemistry, and biogeochemical traits. Particularly, nutrient cycles (e.g., C, N, P, and S) and soil biota are very sensitive to disturbances induced by fire (Oliveira et al. 2014). Fire is a factor of recognized ecological important influencing soil abiotic and biotic traits (Libonati et al. 2020). It also affects the vegetation dynamics, and as previously described, vegetation provides both habitat and energy supply for soil organisms (Arruda et al. 2016). Some studies have reported a reduction of 27% and 38% of the woody vegetation from Pantanal and Cerrado ecoregions, respectively (Berlinck et al. 2022).

Fire events are generally associated with high mortality rates. Fire regime does not aim a specific organism above- or belowground soil surface. It just changes the vegetation physiognomy, the trophic structure, and the entire soil food web (Fig. 8.2) by burning the living and dead organic residues (Arruda et al. 2016). The soil

**Fig. 8.2** Postfire effects on plant community composition. Black gaps tend to increase temperature on soil surface and its exposure to sunlight and wind

organic layers, litter material, and the living organisms are quickly burned during a fire event (Manrique-Pineda et al. 2021). Scientific studies have reported after fire events a decrease of 100% and 75% on soil organic layers, litter material, and living biomass in all areas that were affected by fire and in their surrounding areas, respectively (Oliveira et al. 2014; Austin et al. 2019). It opens great black gaps inside the natural ecosystems that act as follows: (i) increasing soil temperature (it disturbs the entire soil functioning and soil food web activity), (ii) opening a window of opportunity for the biological invasion process, and (iii) increasing soil exposure to sunlight, wind, and water runoff (Berlinck et al. 2022).

When we talk about fire regime, we are encompassing fire severity, fire frequency, the burn season, and the spatial pattern of the burn in a single expression (Cochrane 2003; Junk et al. 2006a, b). Considering the soil abiotic and biotic traits, we must consider that the fire can change the following:

1. Nitrogen mineralization and nitrification
2. Soil P and K leaching
3. Soil microbial biomass C
4. Soil organic carbon pools and C cycle
5. Soil erodibility and infiltration ratio
6. Water flows
7. Soil formation rates
8. Species abundance, richness, and diversity

**Table 8.1**  Threshold temperatures for nutrient and biodiversity loss into soil ecosystem

| Soil trait | Threshold temperature (°C) |
|---|---|
| *Abiotic traits* | |
| Ca | 1484 |
| Clay alteration | 460–980 |
| K | 774 |
| Mg | 1962 |
| N | 200 |
| Organic matter | 100–220 |
| P | 774 |
| S | 375 |
| Soil hydrophobicity | 250 |
| Soil structure | 300 |
| *Biotic traits* | |
| Bacteria | 80–120 |
| Fungi | 60–80 |
| Plant roots | 48–54 |

The fate of nutrient loss (e.g., volatilization, convection, leaching, and erosion) during a fire event has been the subject of scientific research in Brazil, particularly near the Amazon Forest (Cochrane 2003; Barlow et al. 2018). However, the biodiversity loss considering the soil living organisms is unclear. There are few studies in this field of science, and a major part of this studies has just considered the soil microbiota abundance and diversity after fire events. These studies have reported a reduction of 85% and 100% on soil microbial biomass C and microbial C respiration after burning, respectively (Mazzetto et al. 2015).

The impact of fire on soil abiotic and biotic traits is dependent on fire severity, soil temperature, and the temperature of oxidation for each element (Barlow et al. 2018). Soil temperature, for example, is dependent on soil thermal conductivity, soil bulk density, and soil moisture content (Mazzetto et al. 2015). Dry soils, sandy soils, and soils with low bulk density are described with lower specific heat and greater thermal conductivity when compared with moist soils, clay soils, and soils with high bulk density (Cochrane 2003). We also must consider that each soil traits present a different range of threshold temperature (°C) and it determines nutrient and biodiversity loss (Table 8.1).

## 8.3   Landslide Influence on Soil Organic Matter Content

The human activities, illegal buildings in predisposed areas, nature of soils (geology, morphology, and topography), and rainfall contribute to the landslide in the tropical zone (Chen and Martin 2002; Kervyn et al. 2015). The landslide is a kind of natural disaster that takes place in both shoulder and backslope position mainly caused by the deforestation, and it disturbs the soil ecosystem which results in dramatic changes on soil's physical, chemical, and biological traits (Michellier et al. 2016; Błońska et al. 2017). The result of the landslide tends to accumulate on footslope position, and it presents deposits with high variability regarding the distribution of soil organic matter (Balegamire et al. 2017; Broeckx et al. 2018; Kubwimana et al. 2021).

At the tropics, the soil organic matter content (which acts as food resource for the entire soil food web) is used as a bioindicator for soil quality and health (Błońska et al. 2016; Emberson et al. 2020). It also gives us an idea of primary production and helps us to determine some soil traits (e.g., physical, chemical, and biological) (Kitutu et al. 2009; Emberson et al. 2020). However, in areas affected by landslide, there is a high mobility of C into the soil profile, where we can find donator (e.g., by losing C) and receptor (e.g., by receiving C) areas. In such conditions, it is important to determine the changes in the amount of soil organic matter on the shoulder, backslope, and footslope within the range of landslides to predict their influence on the processes of soil cover and vegetation restoration (Mandal 2012; Maes et al. 2017).

In the tropics, the number of landslides has increased at over 26% in the last 5 years because of the urban growth, deforestation, and huge water deposits in predisposed areas (Shiels and Walker 2013; Reichenbach et al. 2018). This is a significant problem in the tropical zone (Mugagga et al. 2012). Previous studies have reported the variability of soil abiotic traits because of the landslides (Shiels et al. 2006), which rare studies considering the variability of soil biotic traits (Błońska et al. 2017; Lacroix et al. 2020). In such conditions, the soil biotic traits have not been used in assessing the restoration process of both soil and vegetation on an affected area by a landslide (Che et al. 2011; Walker and Shiels 2013). In most of the studies, the authors have just focused on the relationship of soil abiotic traits with the landslide occurrence (acting as a prediction for possible landslides) (Wilcke et al. 2003; van Westen et al. 2008).

We must consider the strong distinctiveness from soil properties of the landslide niche in comparison to the footslope area (receptor zone) and the edge area (Mugagga et al. 2012; Bizimana and Sönmez 2015). Some studies have described this distinctiveness as the results of soil texture, soil structure, enzymatic activity, microbial biomass C, and the content of soil organic matter (Guzzetti et al. 2012; Shiels and Walker 2013). This last one is responsible for the saucerful initial stage of soil restoration (Chen and Martin 2002; Sidle and Bogaard 2016). Soil organic matter plays a key role in early soil restoration, and re-establishment of soil functions on disturbed areas (Mandal 2012; Dewitte et al. 2021).

## 8.4   Impacts of Hurricanes on N:P Stoichiometry

Aboveground N and P contents within the tissue of primary producers may reflect plant nutrition (e.g., if they are showing deficiency or sufficiency level) that in turn determines net primary productivity (NPP) at the tropics (Townsend et al. 2007). In tropical forests, for example, their NPP represents a third of the whole terrestrial

NPP (Cramer et al. 1999), and variations in the plant N and P contents and their relationship with nutrient availability are difficult to interpret in affected tropical forests by hurricanes (Dewar 1996). Tropical forests present a wide variety in aboveground stoichiometry, and this variability depends on sampling strategy, plant traits, soil variation, natural disasters, and both horizontal and vertical distribution (Li et al. 2020).

Plant community changes through functional redundancy (e.g., palm trees instead broadleaf trees) in response to the natural disasters can alter stoichiometry (Yang et al. 2020). Given the aboveground N and P contents in palm and broadleaf trees (Augusto et al. 2015), a pronounced increase in palm tree abundance through functional redundancy could theoretically reduce landscape aboveground N and P contents, since broadleaf trees show higher aboveground N and P contents when compared to palm trees. With the more frequent hurricanes expected with climate change (Massmann et al. 2022), palm trees may become more abundant in affected tropical ecoregions (Townsend et al. 2007). Plant community changes have been described in disturbed conditions, and they are related to reduce tropical forest productivity across abiotic and biotic conditions (Paudel et al. 2015), with a corresponding turnover in some plant functional traits (Hooper et al. 2005).

We must expect that, with the diversity of tropical forests, community turnover as influenced by hurricanes could determine aboveground stoichiometry with soil conditions (Cusack et al. 2016). Rather than denoting functional redundancy, and reduced NPP, as has long been hypothesized, aboveground stoichiometry differences could be another bioindicator by which the diversity of tropical forests is constantly changing as influenced by hurricanes and climate change (Silver et al. 1996; Dale et al. 2001; Laurance and Curran 2008; Stegen et al. 2009). It is plausible that, with theses shifts on plant diversity and NPP as affected by hurricanes, the same shifts must occur in soil biota diversity across a wide range of soil conditions (Dale et al. 2001; Li et al. 2020).

Many studies have hypothesized that soil biota diversity (especially ecosystem engineers' diversity) is optimized across natural ecosystems in response to plant diversity (Augusto et al. 2015). Other studies have theorized that aboveground stoichiometry changes as influenced by hurricanes in affected broadleaf forests (Yang et al. 2020), but studies considering the influence of hurricanes on soil biota assemblage are rare. Some theories have suggested that soil biota diversity is positively correlated ($r^2 = 0.81, p < 0.001$) to plant diversity due to habitat and food supply and strongly linked with functional redundancy and negative turnover in tropical forests (Li et al. 2020). This has also been observed in agricultural areas, grasslands, and invaded ecosystems at the tropics (Dewar 1996).

## 8.5 Conclusion

In tropical ecosystem, there are several types of natural disasters with specific effects on soil abiotic and biotic traits. Understanding each natural disaster (e.g., fire, landslide, and hurricanes), their main impacts at the tropics and their influence on soil ecosystem are a hard task and a great challenge for scientists. Impacted areas have difficult access, and sometimes they are close (e.g., especially after the events) for researchers to collect sample through their instability and life risk. It is also hard to predict where the natural disaster will occur, thus being hard to create a timeline pre- and post-event. Using long-term experiments on impacted areas, we can define those natural disasters have strong influence on soil abiotic and biotic traits at tropical ecosystem. We must consider that the natural disasters act disrupting and disturbing both habitat and food resource, which must affect the trophic levels instantly. Nevertheless, we must try to understand nutrient stoichiometry and the soil organisms' community in such areas. Finally, the tremendous variety of soil types at the tropics create a necessity for soil ecologist to recognize the variation plus the influence of natural disasters. Studies considering a timeline after the natural disasters accompanied for a detailed soil characterization are welcome. Especially, if the student is considering both spatial and temporal variation into the impacted area. In view of this, it is important that soil ecologist must consider both soil biota (e.g., identifying taxonomic levels, functional groups, and their behavior) and soil ecosystem characterization (e.g., by providing information about physical and chemical properties).

# References

Arruda WS, Oldeland J, Paranhos Filho AC, Pott A, Cunha NL, Ishii IH, Damasceno GA Jr (2016) Inundation and fire shape the structure of riparian forest in the Pantanal. Brazil PlosOne. https://doi.org/10.1371/journal.pone.0156825

Augusto L, De Schrijver A, Vesterdal L, Smolander A, Prescott C, Ranger J (2015) Influences of evergreen gymnosperm and deciduous angiosperm tree species on the functioning of temperate and boreal forests. Biol Rev 90:444–466. https://doi.org/10.1111/brv.12119

Austin KG, Schwantes A, Gu Y, Kasibhatla PS (2019) What causes deforestation in Indonesia? Environ Res Lett 14:e024007. https://doi.org/10.1088/1748-9326/aaf6db/meta

Balegamire C, Michellier C, Muhigwa JB, Delvaux D, Imani G, Dewitte O (2017) Vulnerability of buildings exposed to landslides: A spatio-temporal assessment in Bukavu (DR Congo). Geo-Eco-Trop 41:263–278

Barlow J, França F, Gardner TA, Hicks CC, Lennox GD, Berenguer E, Castello L, Economo EP et al (2018) The future of hyperdiverse tropical ecosystems. Nature 559:517–526. https://doi.org/10.1038/s41586-018-0301-1

Berlinck CN, Lima LHA, Pereira AMM, Carvalho EAR Jr, Paula RC, Thomas WM, Morato RG (2022) The Pantanal is on fire and only a sustainable agenda can save the largest wetland in the world. Braz J Biol 82. https://doi.org/10.1590/1519-6984-244200

Bieng MAN, Oliveira MS, Roda J-M, Boissière M, Héralt B, Guizol P, Villalobos R, Sist P (2021) Relevance of secondary tropical forest for landscape restoration. Forest Ecol Manage 493(1):e119265. https://doi.org/10.1016/j.foreco.2021.119265

Bizimana H, Sönmez O (2015) Landslide occurrences in the hilly areas of Rwanda, their causes and protection measures. Disaster Sci Eng 1:1–7

Błońska E, Lasota J, Zwydak M, Klamerus-Iwan A, Gołąb J (2016) Restoration of forest soil and vegetation 15 years after landslides in a lower zone of mountains in temperate climates. Ecol Eng 97:503–515

Błońska E, Lasota J, Piaszcyk W, Wiechec M, Klamerus-Iwan A (2017) The effect of land-slide on soil organic carbon stock and biochemical properties of soil. Humic Subst Environ 18:2727–2737. https://doi.org/10.1007/s11368-017-1775-4

Broeckx J, Vanmaercke M, Duchateau R, Poesen J (2018) A data-based landslide susceptibility map of Africa. Earth Sci Rev 185:102–121

Chandra KK, Bhardwaj AK (2015) Incidence of forest fire in India and its effect on terrestrial eco-system dynamics, nutrient and microbial status of soil. Int J Agri Forestry 5(2):69–78. https://doi.org/10.5923/j.ijaf.20150502.01

Che VB, Kervyn M, Ernst GGJ, Trefois P, Ayonghe S, Jacobs P, Van Ranst E, Suh E (2011) Systematic documentation of landslide events in Limbe area (Mt Cameroon Volcano, SW Cameroon): geometry, controlling, and triggering factors. Nat Hazards 59:47–74

Chen C-Y, Martin GR (2002) Soil-structure interaction for landslide stabilizing piles. Comput Geotech 29(5):363–386. https://doi.org/10.1016/S0266-352X(01)00035-0

Cochrane M (2003) Fire science for rainforests. Nature 421:913–919. https://doi.org/10.1038/nature01437

Coyle DR, Nagendra UJ, Taylor MK, Campbell JH, Cunard CE, Joslin AH, Mundepi A, Phillips CA, Callaham MA Jr (2017) Soil fauna responses to natural disturbances, invasive species, and global climate change: current state of the science and a call to action. Soil Biol Biochem 110:116–133. https://doi.org/10.1016/j.soilbio.2017.03.008

Cramer W, Kicklighter DW, Bondeau A, Iii BM, Churkina G, Nemry B, Ruimy A, Schloss AL, Intercomparison TPOTPNM (1999) Comparing global models of terrestrial net primary productivity (NPP): overview and key results. Glob Chang Biol 5:1–15. https://doi.org/10.1046/j.1365-2486.1999.00009.x

Cusack DF, Karpman J, Ashdown D, Cao Q, Ciochina M, Halterman S, Lydon S, Neupane A (2016) Global change effects on humid tropical forests: evidence for biogeochemical and biodiversity shifts at an ecosystem scale. Rev Geophys 54:523–610. https://doi.org/10.1002/2015RG000510

Dale VH, Joyce LA, McNulty S, Neilson RP, Ayres MP, Flannigan MD, Hanson PJ et al (2001) Climate change and forest disturbances: climate change can affect forests by altering the frequency, intensity, duration, and timing of fire, drought, introduced species, insect and pathogen outbreaks, hurricanes, windstorms, ice storms, or landslides. Bioscience 51(9):723–734. https://doi.org/10.1641/0006-3568(2001)051[0723:CCAFD]2.0.CO;2

De Deyn GB, Kooistra L (2021) The role of soils in habitat creation, maintenance and restoration. Phil Trans R Soc B 376:20200170. https://doi.org/10.1098/rstb.2020.0170

Dewar RC (1996) The correlation between plant growth and intercepted radiation: an interpretation in terms of optimal plant nitrogen content. Ann Bot 78(1):1250136. https://doi.org/10.1006/anbo.1996.0104

Dewitte O, Dille A, Depicker A, Kubwimana D, Maki Mateso J-C, Mugaruka Bibentyo T, Uwihirwe J, Monsieurs E (2021) Constraining landslide timing in a data-scarce context: from recent to very old processes in the tropical environment of the North Tanganyika-Kivu Rift region. Landslides 18:161–177

Emberson R, Kirschbaum D, Stanley T (2020) New global characterisation of landslide exposure. Nat Hazards Earth Syst Sci 20:3413–3424

Guzzetti F, Mondini AC, Cardinali M, Fiorucci F, Santangelo M, Chang K-T (2012) Landslide inventory maps: new tools for an old problem. Earth-Sci Rev 112:42–66

Haberle SG, Hope GS, van der Kaars S (2001) Biomass burning in Indonesia and Papua New Guinea: natural and human induced fire events in the fossil record. Paleogeogr Paleoclimatol Paleoecol 171(3–4):259–268. https://doi.org/10.1016/S0031-0182(01)00248-6

Hooper DU, Chapin FS, III Ewel JJ, Hector A, Inchausti P, Lavorel S, Lawton JH, et al. (2005) Effects of biodiversity on ecosystem functioning: a consensus of current knowledge. Ecol Monogr 75:3–35. https://doi.org/10.1890/04-0922

Jacobs L, Dewitte O, Poesen J, Maes J, Mertens K, Sekajugo J, Kervyn M (2017) Landslide characteristics and spatial distribution in the Rwenzori Mountains. Uganda J Afr Earth Sci 134:917–930

Junk WJ, Cunha CN, Wantzen KM, Petermann P, Strüssmann C, Marques MI, Adis J (2006a) Biodiversity and its conservation in the Pantanal of Mato Grosso, Brazil. Aquat Sci 68:278–309. https://doi.org/10.1007/s00027-006-0851-4

Junk WJ, Brown M, Campbell IC, Finlayson M, Gopai B, Ramberg L, Warner BG (2006b) The comparative biodiversity of seven globally important wetlands: a synthesis. Aquat Sci 68:400–414. https://doi.org/10.1007/s00027-006-0856-z

Keefer DK (1994) The importance of earthquake-induced landslides to long-term slope erosion and slope-failure hazards in seismically active regions. Geomorphology 10:265–284

Kervyn M, Jacobs L, Maes J, Che VB, de Hontheim A, Dewitte O, Isabirye M, Sekajugo J, Kabaseke C, Poesen J et al (2015) Landslide resilience in Equatorial Africa: moving beyond problem identification! Belgeo 1:1–22

Kitutu MG, Muwanga A, Poesen J, Deckers JA (2009) Influence of soil properties on landslide occurrences in Bududa district, Eastern Uganda. Afr J Agric Res 4:611–620

Kogan F (2019) Monitoring drought from space and food security. In: Remote sensing for food security. Sustainable development goals series. Springer, Cham. https://doi.org/10.1007/978-3-319-96256-6_5

Kubwimana D, Brahim LA, Nkurunziza P, Dille A, Depicker A, Nahimana L, Abdelouafi A, Dewitte O (2021) Characteristics and distribution of landslides in the populated hillslopes of Bujumbura, Burundi. Geosciences 11:259. https://doi.org/10.3390/geosciences11060259

Lacroix P, Handwerger AL, Bièvre G (2020) Life and death of slow-moving landslides. Nat Rev Earth Environ 1:404–419

Laurance WF, Curran TJ (2008) Impacts of wind disturbance on fragmented tropical forests: A review and synthesis. Austral Ecol 33:399–408. https://doi.org/10.1111/j.1442-9993.2008.01895.x

Li FY, Cy Y, Yuan ZQ et al (2020) Bioavailable phosphorus distribution in alpine meadow soil is affected by topography in the Tian Shan Mountains. J Mt Sci 17:410–422. https://doi.org/10.1007/s11629-019-5705-3

Libonati R, DaCamara CC, Peres LF, Carvalho LAS, Garcia LC (2020) Rescue Brazil's burning Pantanal Wetlands. Nature 588:217–219. https://doi.org/10.1038/d41586-020-03464-1

Lindenmayer D, Thorn S, Banks S (2017) Please do not disturb ecosystems further. Nat Ecol Evolution 1:e0031. https://doi.org/10.1038/s41559-016-0031

Maes J, Kervyn M, de Hontheim A, Dewitte O, Jacobs L, Mertens K, Vanmaercke M, Vranken L, Poesen J (2017) Landslide risk reduction measures: A review of practices and challenges for the tropics. Prog Phys Geogr Earth Environ 41:191–221

Mandal TN (2012) Restoration in soil and plant properties in landslide damaged forest ecosystems. Nepal J Biosci 2:40–45

Manrique-Pineda DA, Souza EB, Paranhos Filho AC, Encina CCC, Damasceno GA Jr (2021) Fire, flood and monodominance of *Tabebuia aurea* in Pantanal. For Ecol Manag 479(1):e118599. https://doi.org/10.1016/j.foreco.2020.118599

Massmann A, Cavaleri MA, Oberbauer SF, Olivas PC, Porder S (2022) Foliar stoichiometry is marginally sensitive to soil phosphorus across a lowland tropical rainforest. Ecosystems 25:61–74. https://doi.org/10.1007/s10021-021-00640-w

Mazzetto AM, Feigl BJ, Cerri CEP, Cerri CC (2015) Comparing how land use change impacts soil microbial catabolic respiration in Southwestern Amazon. Braz J Microbiol 47(1). https://doi.org/10.1016/j.bjm.2015.11.025

Michellier C, Pigeon P, Kervyn F, Wolff E (2016) Contextualizing vulnerability assessment: A support to geo-risk management in central Africa. Nat Hazards 82:27–42

Morgado RG, Loureiro S, González-Alcaraz MN (2018) Changes in soil ecosystem structure and functions due to soil contamination. In: Cachada A, Rocha-Santos T (eds) Duarte AC. Academic Press, Soil Pollution, pp 59–87. https://doi.org/10.1016/B978-0-12-849873-6.00003-0

Mugagga F, Kakembo V, Buyinza M (2012) A characterization of the physical properties of soil and the implications for landslide occurrence on the slopes of mount Elgon, eastern Uganda. Nat Hazards 60:1113–1131

Nunes MH, Jucker T, Riutta T et al (2021) Recovery of logged forest fragments in a human-modified tropical landscape during the 2015-16 El Niño. Nat Commun 12:e1526. https://doi.org/10.1038/s41467-020-20811-y

Oliveira MT, Damasceno GA Jr, Pott A, Paranhos Filho AC, Suarez YR, Parolin P (2014) Regeneration of riparian forests of the Brazilian Pantanal under flood and fire influence. For Ecol Manag 331(1):256–263. https://doi.org/10.1016/j.foreco.2014.08.011

Paudel E, Dossa GGO, de Blécourt M, Beckschäfer P, Xu J, Harrison RD (2015) Quantifying the factors affecting leaf litter decomposition across a tropical forest disturbance gradient. Ecosphere 6(12):267. https://doi.org/10.1890/ES15-00112.1

Ponge J-F (2015) The soil as an ecosystem. Biol Fertil Soils 51:645–648. https://doi.org/10.1007/s00374-015-1016-1

Reichenbach P, Rossi M, Malamud BD, Mihir M, Guzzetti F (2018) A review of statistically-based landslide susceptibility models. Earth-Sci Rev 180:60–91

Shiels AB, Walker LR (2013) Landslides cause spatial and temporal gradients of multiple scales in the Laquillo Mountains of Puerto Rico. Ecol Bull 54:211–221

Shiels AB, Walker LR, Thompson DB (2006) Organic matter inputs variable resource patches on Puerto Rico landslides. Plant Ecol 184:223–236

Sidle RC, Bogaard T (2016) Dynamic earth system and ecological controls of rainfall-initiated landslides. Earth-Sci Rev 159:275–291

Silver WL, Brown S, Lugo AE (1996) Effects of changes in biodiversity on ecosystem function in tropical forests. Conserv Biol 10:17–24. https://doi.org/10.1046/j.1523-1739.1996.10010017.x

Slimane M, El-hafid N (2021) Microbiome response under heavy metal stress. In: Sharma A, Cerdà A (eds) Kumar V. Elsevier, Heavy metal in the environment, pp 39–56. https://doi.org/10.1016/B978-0-12-821656-9.00003-1

Stegen JC, Swenson NG, Valencia R, Enquist BJ, Thompson J (2009) Above-ground forest biomass is not consistently related to wood density in tropical forests. Glob Ecol Biogeogr 18:617–625. https://doi.org/10.1111/j.1466-8238.2009.00471.x

Townsend AR, Cleveland CC, Asner GP, Bustamante MMC (2007) Controls over foliar N:P rations in tropical rain forests. Ecology 88(1):107–118. https://doi.org/10.1890/0012-9658(2007)88[107:COFNRI]2.0.CO;2

van Westen C, Castellanos E, Kuriakose SL (2008) Spatial data for landslide susceptibility, hazard, and vulnerability assessment: an overview. Eng Geol 102:112–131

Vogel H-J, Balseiro-Romero M, Kravchenko A, Otten W, Pot V, Schlüter S, Weller U, Baveye PC (2021) A holistic perspective on soil architecture is needed as a key to soil functions. Eur J Soil Sci 73:e13152. https://doi.org/10.1111/ejss.13152

Walker LR, Shiels AB (2013) Introduction for landslide ecology. USDA National Wildlife Research Center – Staff Publications. Paper 1642

Wilcke W, Valladarez H, Stoyan R, Yasin S, Valarezo C, Zech W (2003) Soil properties on a chronosequence of landslides in montane rain forest, Ecuador. Catena 53:79–95

Yang B, Qi K, Bhusal DR, Huang J, Chen W, Wu Q, Hussain A, Pang X (2020) Soil microbial community and enzymatic activity in soil particle-size fractions of spruce plantation and secondary birch forest. Eur J Soil Biol 99:e103196. https://doi.org/10.1016/j.ejsobi.2020.103196

Yunus AP, Fan X, Subramanian SS, Jie D, Xu Q (2021) Unraveling the drivers of intensified landslide regimes in Western Ghats. India Sci Total Environ 770:e145357

# Appendices: Supplementary Material

## Appendix A: Pathway to Classify Soil Organisms Proposed by Souza (2021)

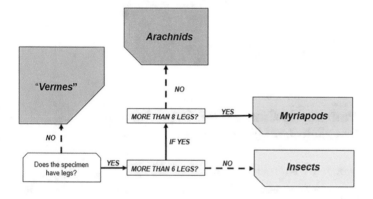

## Appendix B: Pathway to Classify Soil Organisms into the Branch of "Vermes" (Souza 2021)

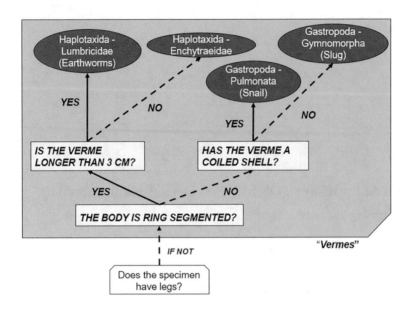

## Appendix C: Pathway to Classify Soil Organisms into the Branch of "Arachnids" (Souza 2021)

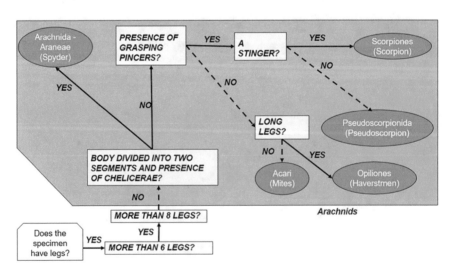

# Appendix D: Pathway to Classify Soil Organisms into the Branch of "Myriapods" (Souza 2021)

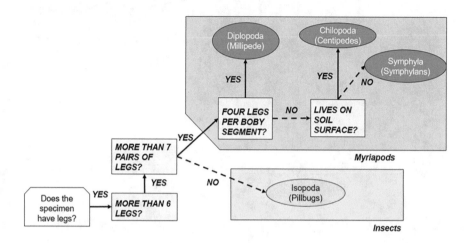

# Appendix E: Pathway to Classify Soil Organisms into the Branch of "Insects" (Souza 2021)

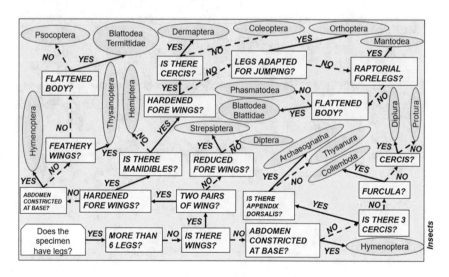

# Reference

1. Souza TAF (2021) Pathways to classify soil organisms. UFSC: PPGEAN, Santa Catarina, Brazil. 56p. ISBN: 978-85-455046-2-7

Printed in the United States
by Baker & Taylor Publisher Services